Phantasmagoria

The New Frontier

The New Frontier

Phantasmagoria

By Raymond Bolen

This book is dedicated to everyone who believed that this dream of mine would come true and those who supported me during the endless hours of writing this novel. Thank you all so much!

Cover photo courtesy of NASA

And NASA/Ames/SETI Institute/JPL-Caltech

All references to companies, organizations, or nations are solely for the purpose of storytelling. In no way is this work supposed to disrepute or discredit the name of such entity.

Printed by CreateSpace, an Amazon.com Company

Phantasmagoria

Prologue

It is the destiny of all stars to die. The time for the death of a star cannot be determined by any known scientific instrument. When the time comes for a star to die, it dies. Just like that. The process of star death is complicated with a star going through several stages before it is finally dead. As they die, many stars explode. These are massive explosions, fueled by trillions and trillions of tons of hydrogen fuel.

It was an explosion such as this that occurred several thousand years ago setting the stage for this story. A star, not too distant from our own sun by stellar distances, exploded sending burning gasses and radioactive debris far beyond its original gravitational field. All of the planets orbiting this sun were instantly destroyed. All of them that is, except for one, an Earth-sized planet with an unusually thick atmosphere and thick water canopy. The water canopy and much of the atmosphere were destroyed, but they served to protect the planet itself from sustaining too much damage.

Phantasmagoria

Without a star to orbit, the planet began a long, slow trek across the universe in search of another star. The remaining atmosphere froze solid, preserving the planet for some future star to claim and cherish. This time-suspended planet faced many risks on its pilgrimage. The planet was wandering aimlessly like a boat without a rudder, setting itself up to go crashing into a star or other planet. It could end up missing all of the stars in the local star cluster in which case it would head out to deep space and run the risk of getting caught in the gravity trap of a black hole. Bravely facing these challenges, the small sphere of rock and dirt sought out a new home.

Chapter 1

My boss, Todd Reeves, was storming around the interior of the small store shouting in rage. "When you FINALLY finish working on the Piper, there is a GH-04 with some stuck bearings for you to work on!" Todd screeched. "You had better have them done by the time I get back or you're through!" The door to the little shop opened letting the everyday noises of the town in, and the cantankerous shop owner out.

"Whew, Lawrence," I said, looking at my assistant and friend who was seated to my left. "How long do you think it will be till he gets back?"

"I don't know," replied Lawrence as he picked through a jumbled mass of tools for a screwdriver. "But we had better try to get as much done as possible. Whew, Todd sure messed this toolbox up." Without saying anything more, we both bent our heads down over the model aircraft we were repairing and continued working. Todd Reeves is far from the boss most employees dream of. When he gets in one of his moods… well, you'd better just stand back and give him what he wants. Of course, that is quite difficult to do with my personality. I am a very strong-willed person and I do not like to be told what to do by a guy with only a high school diploma, considering that I have a Master's Degree. I have worked for Todd at his hobby shop for three years and can testify that 'the old wart' would probably flunk out of even the developmental courses at the local community college. The only reason that he is still in business is that he made some good marketing decisions after the economic crash of 2020.

"Can you pass me a Phillips #2?" Lawrence asked me. I slid the requested screwdriver across our long repair table/sales counter. His thanks were muffled by the ringing of the telephone. Being the closest to the phone, I reached over and picked it up.

"Reeves Hobby Shop," I said into the receiver, "this is Flynn Carson."

"Flynn, this is Jerry." Jerry Gonzales had been my best friend since the first grade.

"Hi Jerry, what's on your mind?" I asked, wondering why he was calling during business hours.

"Can you meet me at Burger Corner after work tonight? We've got to talk about the wedding."

"That sounds fine." I replied. "Will 7 o'clock work?"

"Terrific."

I hung up smiling at the future groom's nervousness. In two weeks Jerry would be married to his college sweetheart, Cheryl Sanchez, and I was to be the best man. Jerry was extremely eager and excited about the upcoming event. However, he was also very nervous and he was always talking about the 'what-ifs'. That was starting to get on my nerves.

"Does Jerry want to talk about his wedding again?" Lawrence asked with a grin.

"Yes," I said chuckling. "We should probably give him a break though. I suppose that getting married could make a guy quite nervous."

Lawrence thought about that for a moment. "Nah, he is still being a worrier. At least we will have something to tease him about later."

I grinned and returned to my work. I put the landing gear back on the model Piper Cub and carried it over to the table that held the repaired models. As I returned to the counter carrying the GH-04 helicopter, I took a quick glance around the small room. The shelves in the old hobby shop were only partially full with cheap, discount models and kits, many of them factory seconds. With the world economy in the dump, people could only afford hobbies if they were cheap and Reeve's Hobby Shop had survived the collapse by selling only the cheapest plastic model kits. Rumor had it that Todd had also sold inexpensive surveillance drones to the law enforcement in exchange for tax reprieves. It was the increased taxes that sunk the United States into poverty along with the rest of the world. But that was back in 2020, when I was only ten years old. By 2028, when I graduated from high school, the economy had improved enough for my parents to be able to pay for my college education. Now in 2038, everyone is still practically broke, but the end of the tunnel is in sight. Commerce is starting to pick up in the United States and that is what a capitalist economy thrives on.

I replaced the frozen bearings in the helicopter's drive system and put it back together. Lawrence had finished assembling the airplane he had been working on and was dozing in his chair. I checked my watch. 2:45. Todd had been gone for an hour and a half and it was time for me to head to the airfield where I gave flight lessons part-time. I grabbed my car keys and walked out into the warm Florida sunshine. Being the middle of the afternoon, the traffic in my little town was minimal. I climbed into my old Ford pickup and backed into the street. It was a hot day in late June and I was thankful for the air conditioning.

Phantasmagoria

∞

"How did the lessons go?" Jerry asked with a grin, noticing the grim look on my face. I sat down in the booth seat across from him before replying.

"Ryan Butler was late as usual, and then my left aileron was a bit sticky. By the time we got that taken care of, the wind had picked up so that the turbulence was awful."

"I'm sorry." Jerry said with feeling. My students are notorious for being late. Thankfully my family and friends understand when I show up someplace a little late or nauseous. A pretty waitress approached our table with an order pad.

"What can I get for you gentlemen today?"

"I'll take a large cheeseburger and fries," Jerry said, "and an extra pickle."

"And what can I get for you?" She asked me, smiling.

"I guess I'll take the same," I said with little enthusiasm.

"Does Ryan need more practice handling wind?" Jerry asked. I nodded and started to sip the glass of water that Burger Corner always kept on their tables. Then I leaned back against my seat and closed my eyes hoping that the nausea would pass. I usually didn't get motion sickness, but this afternoon was an exception. Jerry reached over and turned on the mini TV at the end of the table. As the TV came to life, I could hear a man talking with the enthusiasm and energy of a lawyer whose client is about to be sent to death row. I opened my eyes and saw a man in his late sixties standing behind a podium with an American flag in the background. That would be President Watson. He was obviously losing a debate against his political rival, Andrew Lee.

Andrew was a man in his early fifties and highly favored in the polls. Since the collapse, all presidents had been one-term presidents on account of their lack of ability to reverse things. So far, Andrew looked to be the best option to replace President Watson. He was smart, handsome, and very successful. Everybody loved him.

"Think the election will be a landslide this time?" Jerry asked in an obvious attempt to distract me from my symptoms. I looked at my best friend a moment before answering.

"Lee will win, but there are a lot of people who like Watson. He has done more than any of the previous presidents."

"But you know politics." Jerry said, trying to defend his idea of a landslide. "Lee won't let the people forget their empty pocketbooks. Those are still the same since the last election"

"He won't let them forget that, but the people will also remember that Watson kept us out of a war with India."

"You know that war was a bunch of nothing." Jerry said, starting to get slightly aggravated.

I smiled, knowing that Jerry was starting to give in to my point of view and said, "But not many people believe that. They think that a couple of foolish people in the White House almost started a war with the world's only superpower and Watson stopped them."

"You know India isn't a superpower! The days of superpowers are over, man. After the United States debt default and the Russian plague, China almost had a world monopoly until their civil war. Now the world is equally poor with no opportunity to get ahead."

"When the sand is sifted, the capitalist countries will always come out on top," I said. "We just need the right president to help us out."

"Lee is that president and the people know it!" Jerry said with satisfaction. "I rest my case."

Phantasmagoria

I stared at Jerry for a moment and then shook my head. Jerry just smirked. About that time the waitress approached carrying our meals. She set them on the table, told us to enjoy and left to tend some more customers that had just arrived. Jerry and I ate in silence while watching the debate on the TV. The candidates were debating foreign policy and Lee was simply frying the President.

The debate ended a few minutes later with Watson making a last-try insult and it falling flat. I ate my last fry and pulled a couple of dollars out of my pocket to pay. We each left a dollar on the table for a tip and prepared to leave. Just as we were standing up, the words 'Breaking News' flashed across the screen in bright letters. An announcer came on, talking rapidly. Above her left shoulder was a screen displaying an image of some stars with a very bright star in the center. The headline read: Astronomers spot new comet on collision course with earth.

"Turn it up. I can't hear." Jerry said.

We sat back down and I turned the volume up a little bit.

The news caster was saying, "…astronomer at the Colorado observatory said that the comet is entering the solar system at a trajectory almost even with the orbits of the major planets. It is currently almost on the other side of the sun, but by the time it approaches Earth's orbit, the Earth will be on the comet's side. Advanced calculations show that the Earth and the comet will collide on the sixth of January, 2039. As of this time, no information has been released about the size of the comet or the expected damage the planet will receive."

The reporter stopped talking and Jerry turned the set off.

"I don't believe that any more than I believe in zombies," Jerry said with disgust.

"Well, you remember the incident in Dallas, don't you?" I replied, jokingly.

The New Frontier

"That was a government conspiracy and I think the same is true with this."

"I don't know, but astronomers are usually pretty trustworthy."

"But if the government is in on it, then they can get anyone to say anything."

We paid our bill and headed out to the parking lot. As soon as I got into my truck, I congratulated myself on my success in getting Jerry to momentarily forget his wedding. Lately, it had become almost impossible to be around Jerry without him making some mention of his wedding.

It was about 8:30 at night and the small streets of Hawthorne, Florida were practically deserted. This was very unusual for a Thursday night. Typically, the streets would be crowded with people heading to and from the movie theatres and restaurants. I got to my house on the outskirts of town, and sat down to watch some TV. Almost all 500 channels had something about either the comet or the election. Disgusted, I turned it off and went to my bedroom and lay down in bed to muse.

What is such a big deal about a comet? Astronomers have thought many comets and asteroids were going to hit the Earth, but they ended up missing. The Earth was a very small target in a vast expanse of space. A hundred-thousand miles is an extremely large distance and even the moon is much farther away than that. Was it possible for a bunch of guys with telescopes to be able to predict where an object on the other side of the sun would be in six months? Could they make a prediction of such proportions that isn't even thirty-thousand miles off? Even if it is accurate, why is it all over the media so soon? For crying out loud, nobody even knows how big the thing is! These thoughts crowded into my head like fans in a football stadium and would not let me rest. Nevertheless, no matter how much I tried to reason that the comet was an error, a dull dread filled the bottom of my stomach. Those guys weren't paid 80 grand a year to be wrong.

Phantasmagoria

I lay on my bed thinking about what would happen if a comet were to hit the Earth. I knew comets are made of ice and rock. If the thing is big enough for astronomers to see it on the other side of the solar system, then it must be pretty big. It would be big enough to break the planet into pieces. And it might even be closer to Earth than they originally thought! I began to hear a low whistling sound that resembled a projectile flying through the air. It grew louder and louder. I instantly knew it was the comet! It was coming through the atmosphere and was going to hit nearby! The noise grew unbearably loud. I ran outside and looked into the sky. There was the comet, streaking across the sky. It filled the entire sky with white flames! Then it crashed into the ground. A shock bigger than any I ever felt rattled my bones. I could feel the ground erupt beneath me as the planet turned to fragments. I could feel the air leave my lungs as I shot out of the non-existent atmosphere into space. Trees, rocks, and dirt flew past me in pure weightlessness. I gasped in the vacuum as my lungs struggled for something to inhale. Then everything went blank.

∞

"Are you still having the nightmares?" asked a sweet voice. I looked my girlfriend across the table and grinned.

"I told you it was just that one night."

"I know, but it's been a month and I can see that you are still worried about the comet."

"No, I'm just upset that there is nothing we can do except sit around and die."

"Do you think they will elect a new president?"

"Of course they will. Life still has to go on, comet or no comet."

"The news reports say it is as big as Earth."

"I know." I looked out of the pizza place and across the street to the shopping center. The place looked almost deserted. Since the comet was first seen, many people had almost completely stopped buying things that were not needed for everyday living. They only spent their money on food, movies, alcohol, and vacations. Protestors of this laziness stood outside the grocery store and waved signs telling people not to trust the government and continue life as normal. They reasoned that even if the end of the world was imminent, we should make our last days meaningful instead of wasting them partying. "Debra, do you honestly believe in this comet?"

"Yes, the scientists who found this thing are very smart."

"They are, but it is very far away."

"I know. I guess we will just have to wait and see."

Chapter 2

"Lee has won the election," said the reporter. "Andrew Lee is the 50th president of the United States of America."

I looked at the results while sitting on the couch in my living room, watching the large flat-screen TV, and eating some gas station pizza. It was not a bad life for a bachelor. Lee had defeated Watson by 16 electoral votes, far from the landslide that Jerry had predicted. I had to call him and rub it in.

"Jerry?"

"Don't say anything." He snapped in his Mexican accent.

"I told you so."

I heard him grunt on the other end.

"When are you and Debra getting married?"

"You're changing the subject. Sure sign of defeat."

"Shut up. You had better marry that girl before she slips through your hands. It's not every day a fellow like you gets a girl like her."

"Well Mr. Expert, I proposed last night."

"You did? Man! Let me be the first to congratulate you!"

"You are a day late and a dollar short. Lawrence beat you to it."

"Shucks. Did you hear the update on the comet?"

"There you go changing the subject again."

"No, I am serious. It was affected by the gravity of Mars and it isn't going to hit Earth after all."

"Are you kidding? Why is this not all over the news?"

"It's because a new president was just elected. If you would stop sleeping in till nine in the morning you might see some of this stuff before it becomes old news. You will see plenty more about it in a little while."

I looked over at the TV screen.

"Jerry, it's not even a comet."

"What do you mean?"

"I am looking at the TV right now. They say it is a planet."

"And it had to pass Mars before they figured that out?"

"Jerry, just go to your own TV and watch it. We might have more of our own questions answered if we hang up."

"Bye, *click.*"

I turned up the volume on the TV.

Phantasmagoria

"The comet which caused a global commotion earlier this year has turned out to be a planet from an unknown system. Current reports say it will miss the Earth entirely and form a stable orbit between Earth and Venus. The previous calculation that put the planet on a collision course with Earth was a simple computer glitch that has been fixed…

A lot of security that is, I thought. *If they can make one error, then they can make a dozen.*

"The planet has been named 'Phantasmagoria'. This name was chosen because of the multiple visual changes it went through as it warmed up and its atmosphere melted and boiled. Phantasmagoria is a planet almost the exact size of Earth. Our latest reports say it has an atmosphere very similar to Earth's and is capable of supporting life. Of course, any life that was once on it would have been killed off long ago when it left its home system and froze. Scientists still have no clue as to Phantasmagoria's origin or how it left its first star, assuming it had a star. If Phantasmagoria is found to never have orbited a star, then it may yield scientists an important clue as to the formation of our universe…"

The reporter continued to drone on and on about what scientists thought about the planet. It seemed as though everyone and his uncle had an opinion as to where the planet may have come from or what it may be like. There were even several theories about how aliens might use the planet to travel from system to system, checking on their subordinates. Another small group was blaming the government for hiding the truth about the planet's trajectory and the fact that it was a planet. After all, they reasoned, why would no one see it until it was within the orbit of Jupiter, and how could they tell it was a comet at first? Defenders of the astronomers declared that they had tracked the planet for a couple of years as it entered the solar system, but they thought it was nothing to worry about until a report showed its possible crash with Earth. They proved this by showing some old newspaper articles about an 'invader from space'.

I watched for a while before calling Debra.

The New Frontier

"Hi Sweetie, how are you?"

"I am just fine. Have you been watching the news?"

"Yes I have. For once they actually have something interesting on."

"That's true. What do you think of the election?"

"I think that the economy is going to pick up with the new reports on Phantasmagoria and that Lee is going to take credit for it. He will be the first two-term president that we have had in years."

"Why do you think the economy will pick up?"

"It will pick up because people now have lives to live. The world is not going to end soon. They now have to maintain houses and yards, they have to stock shelves, they have to buy stuff. When people buy stuff in a capitalist society, the economy flourishes. In the past, people just have not had to buy enough stuff to get all of the way out of the depression. There is about to be a huge buying spree and that will push America over the edge and into prosperity."

"That is an interesting theory. I hope you are right. We could sure use some economic growth"

"That we could," I agreed. "I'll talk to you later"

After I hung up, I sat there pondering another possible conspiracy theory I had not heard mentioned on the news. Was it possible that the comet or planet or whatever it was supposed to be was still on a collision course with Earth and the world leaders had decided to lie to prevent panic in the remaining months of civilization? That was possible, but why would they allow the world to panic for three months? I grabbed a notepad and pencil and started listing the things I knew about the publicity of Phantasmagoria.

It was first brought to public attention at the end of June during a time when the presidential race was very hot. This planet or whatever it is was first seen a few years ago, but did not make major headlines. It was not until recently that someone noticed, or thought they noticed, that is was going to hit Earth. Maybe someone was making a model of the solar system and they fast-forwarded a couple of months and saw the impending danger. They then showed their supervisor, who then showed his, and so up the line it went till the information was released to the media. The one thing that kept screaming 'conspiracy' was the fact that it took them three months to find the error. With the entire world checking and rechecking calculations one would figure that the error would be found out almost immediately.

Another point I considered was that they had mistook a planet the size of Earth for a comet. I had read several articles about the accuracy of Hubble 2 and how it could take accurate images of planets in the Centauri system. If it could take pictures of a planet that is four light-years away, shouldn't it be able to tell the difference between a comet and a planet in our own system? I stepped into my bedroom and returned with my laptop. After a couple of minutes searching the internet I found that many others were asking the same questions I was. It almost looked as if a bigger ruckus was created when people found out that there would not be a collision than when they though there would be.

∞

Two months have passed since the media revealed the error in the original calculations of Phantasmagoria's trajectory. The conspiracy theory following has diminished to a shrinking minority. For the past two weeks, Phantasmagoria has been visible in the night sky and a few people say they have spotted it in the daytime. The media says that it will pass by the Earth at a distance of half-a-million miles. They say that the Earth's gravity will act as a brake to Phantasmagoria and allow it to fall into a slightly elliptical orbit around the sun. Scientists have been arguing constantly about the possible effects it will have on the Earth. Some say the horizontal gravity will topple buildings and slow down the Earth's rotation causing longer days. Others say that it will only increase the tides. I guess the only thing we can do it sit back and wait.

∞

"Flynn, what do you think it is like on Phantasmagoria?"

I looked at the beautiful girl lying on the blanket beside me.

"I don't know Debra, but maybe someday we will send astronauts to visit it and they will tell us."

"It is so beautiful," she said. "I think if all of the land on Earth were desert, then Earth would look like Phantasmagoria."

I looked at the newest addition to the solar system with the awe and amazement of one who has just seen a tree-covered mountain range for the first time. Being just over half of a million miles away, Phantasmagoria was a little over four times the size of a full moon. Its oxygen-rich atmosphere was highly visible, reflecting the sun's light back to Earth brighter than any full moon ever did. Phantasmagoria looked all alone in the depths of space, its shine blocking out all other stars. Much of the surface was covered with clouds making the planet resemble Earth, except for one very noticeable characteristic. Its vast continents were a Florida beach sandy white. Not a Sahara yellow, but Emerald Coast white. The oceans which covered sixty percent of the planet were a deep blue like those of Earth, but when contrasted with the land, they gave the planet a very ghost-like or dream-like appearance in addition to its awesome beauty. Phantasmagoria… That was truly a fitting name for such a planet.

I sat there in the cool night air wondering how many people had dreamt of such an occurrence happening in their lifetime. Sure, many people had dreamt of being around to see a new technology unfold, or a war end, or a change in civilization, but who had dreamed of seeing a new planet enter the solar system and join our cozy little family of worlds? I concluded that I was a very special person to have the chance to witness such an event. It also occurred

Phantasmagoria

to me that this was more than just history unfolding, this was nature revealing to mankind some more of its amazing laws. Generations from now, people would accept Phantasmagoria as a natural planet of the solar system and never question its being here during the creation. As technology improves and civilization collapses from apathy and ignorance, people will assume that Earth has always been the forth planet from the sun. Who knows, but perhaps one of the nine planets we have thought always belonged to the sun is in fact another hitchhiker, riding from one system to another as the suns burn out and cast away their outer satellites. It is very possible that such knowledge was lost during the Great Flood or the Dark Ages.

A gust of cool air temporarily brought my thoughts back to reality and my fiancée lying beside me. I looked at Debra lying there in the bright light of Phantasmagoria and thought about my future with her and my future with the rest of the world. It would not do for me to live out the rest of my days fixing the bearings on model helicopters or teaching high school students how to fly a bundle of tubes and fabric. What was the purpose of going college and getting a degree in mechanical engineering if I was not going to use it for something larger? It would have to be something larger than me, larger than Debra or my family and friends. As a kid I always dreamed of changing mankind for the better, but during college, well, bills hit and there I was at Reeve's Hobby Shop. But bills didn't stop others who had changed the world, they sucked in their gut and plowed ahead to face whatever obstacles faced them. I had planned to do the same, but somewhere along the way I lost sight of my goal. My goals... I have never been a person to work without goals. I figure out what I want to do or where I want to go, then I do what it takes to get there.

But how does a person change the world? It must not be something you think about, it must be something you do. It must be like saving someone from a burning house. You never think about what you are doing, you just do it. And we cannot all be heroes. Everybody says they would save someone from a fire, but when the time actually comes you never know who is going to rush into the flames and who is going to freeze in fright. The same must be true about the people who have made great advancements in their field whether it is science, politics, or literature. People like Shakespeare, Newton, and Washington all dramatically changed history, but did any of them deliberately strive to better mankind or did they simply do their daily job with all of their strength and devotion, while hoping that someday everyone would benefit from the effects?

We sat out under the cheery glow of the new planet for a few minutes longer before going inside my house to get something to drink before Debra headed home for the night. I had a lot of serious thinking and self-evaluation to do.

Chapter 3

"I will tell you for the third time Jerry, I am not an astronaut nor will I ever apply for something so crazy."

I hung up the phone and turned to Lawrence. He was trying to appear busy on a model airplane, but I knew he had been actively listening to every word.

"Jerry is trying to get me to apply for the Spacex mission," I said.

"Do you mean the one to Phantasmagoria?"

"Of course I mean the flight to Phantasmagoria," I snapped, annoyed because Lawrence knew exactly what I was talking about. "That mission will be the only spaceflight in a decade."

"What do you have against applying?"

I knew Lawrence had no impulse or sense of adventure, but he was famous trying to convince other people to 'go all out.' In fact, I secretly wanted to apply, but I wanted to be persuaded by Debra or myself, not Jerry. Jerry and I would never allow ourselves to be swayed into something by the other, especially if we had showed opposition to the idea in the first place. 'The Spacex mission' as the public referred to it, was a proposed flight to land six crews of seven people each on Phantasmagoria and return them to Earth after they had explored the surface for a couple of days. The idea was first proposed on the eighth of January while Phantasmagoria was still high on the headlines. Ever since the Mars mission was canceled in 2025 due to decline in public interest, Spacex officials have been debating how to use the Falcon Heavy rockets that had been manufactured for the mission. Now nature had provided a way.

The proposal of a flight to the new planet took the public by surprise. I thought that the idea was terrific. Spacex had been going nowhere financially for the last fifteen years and a flight to Phantasmagoria was just the thing to put them back into business. However, since no country or company had been sending people to space for a while, there were no trained astronauts around. Spacex had started accepting applications from the general public in an attempt to also raise public enthusiasm.

"It isn't that I do not want to apply, Lawrence, but I don't want to apply while Jerry is telling me to. It is kind of a pride thing I guess."

"I see. Would you apply if Debra told you to?"

"I probably would."

Our conversation was ended by Todd Reeves coming in the door from his three-hour lunch break. One look at the boss told me that the rest of the day would be miserable. He slammed the door and steamed over to the cash register, slamming his briefcase on the counter. He opened the cash drawer and fingered the bills.

"We have not had much business I see," he said bitterly.

"No sir, just a guy buying some batteries." Lawrence said timidly.

Todd stormed into the back room while mumbling something about the world not giving him the breaks he deserved. Lawrence looked at me and shrugged.

"I wonder what happened to him." Lawrence pondered.

"He probably ate Korean food. He says that bothers his stomach ulcers."

Lawrence chuckled and we returned to our work. Todd suddenly stormed into the room.

"Flynn, get those plastic models I ordered out onto the shelves! That should have been done yesterday," He shouted. "Why do you two never take any initiative and do anything on your own? I have to walk you two around like babies and tell you to do this, and tell you to do that. If ya'll were in charge of this store, it would fall apart in a month. No, it would fall apart in a week and the bank would foreclose on you in a month. What are you doing standing around Flynn? Get those shelves stocked!"

I felt bad about leaving Lawrence to listen to Todd, but I did not need another invitation to leave. I hustled into the back room and pulled the little plastic models of ships and planes out of the big cardboard box they had been shipped in and ran an armload out to the storefront. Someday when Debra and I had kids, I would help them put models like these together, but I would not give Reeves the business. I would even drive out of town if I needed to, just to buy them from another shop.

When I walked out of the back room, Lawrence was trying to appear busy on a model submarine and Todd was stalking around the shelves, mumbling about how lazy Lawrence and I were. I suppose it had never occurred to Todd that he had told us earlier that the company sent the wrong models and he was going to return them. He would probably remember next week and Lawrence and I would suffer. I knelt down and filled the shelf with the small boxes. It was kind of insane to buy more models when we could barely sell the ones we had, but Reeves would not answer to reason.

I had just but the last model on the shelf when Todd walked over and started yelling something about making Lawrence and I making sure that sales increased by ten percent in February or we would be looking for other jobs. He whirled around and stomped out the door while telling us to make sure we locked the place up when we left.

"I wish I could leave early in the afternoons and give flight lessons like you do," Lawrence said as the door slammed behind our angry boss. "It must be a great way to relax, being up there high above your problems."

"That is true, except for the fact that my student has the flu today." I stood up and almost sprinted to the cashier computer. I sat down on the stool in front of it and started typing.

"What are you doing?"

"I have just about had enough of Todd Reeves." I said angrily. "I am not going to spend my life working for a crab that cares nothing about his business or his employees. If all Todd cares about is his pocketbook, then I am going to let him fill it himself. I am going to spend my days doing something that I enjoy and can be productive at. I am going to send an application to Spacex for whatever positions they have openings for."

"You are definitely qualified for a number of positions."

I glanced over at my friend. He was bent over the submarine tinkering with its insides.

"Lawrence, you are also qualified to work for them. Why don't you apply too?"

He looked up and his face brightened. You mean I wouldn't have to work for the old warthog anymore?"

"You can always apply. What is the worst thing that can happen? If they decline you, you can always put in an application somewhere else."

"That sounds great! Hey, do you think they might give us positions in the astronaut program?"

"I doubt they would do that, but now that they are launching stuff, there will be many other positions open. Then again, anything is possible"

"Why would they not hire us for the astronaut program? You are a pilot, I studied geology in college, and we both spend our time fixing airplanes."

"You have to realize that there will be many others applying for these jobs. Even if we are some of the most qualified, it is possible that our applications will be lost in the stack of paperwork. And we fix model airplanes."

"How come I am being more optimistic than you?"

"Lawrence, it is not a matter of being optimistic, I just do not want you to get your hopes up for nothing. See, I am putting my application in the astronaut program part of their website."

It took Lawrence and me the rest of the afternoon to fill out the online application forms and submit our resumes, but the time passed quickly in our excitement of the possibility of no longer working for Todd Reeves. As I headed home, I thought about what I would tell Debra and Jerry. Jerry would laugh his head off and say 'I told you so', but there was no telling how Debra would take it. Whenever Jerry had mentioned it, she had seemed positive about it, but I supposed she assumed Jerry was joking. However, I knew my buddy better than that. I decided to call Debra first and tell her. I tried her home number, but got her answering machine. I then tried her cell phone.

"Hello."

The New Frontier

"Hi Debra, it's me."

"Hi Flynn, what's up?"

"I just wanted to tell you that I applied for the flight to Phantasmagoria."

"You applied? That's wonderful! Do you think they will select you?" I could tell by her tone of voice that she was worried.

"I have no way of knowing. I feel like I am qualified, but you never know with these guys. There may be thousands of applicants. I was hoping that you would like the idea. Space flight is not too dangerous."

"Oh, I am not worried. After all, statistics say you are in more danger driving home right now than you would be in space," she said slightly sarcastically.

A sports car pulled out right in front of me, causing me to slam on my brakes and swerve into the other lane.

"Thanks a lot," I said.

"Of course if you are selected, I will get real nervous before the launch and probably try to talk you out of it. I do think it would be a good way to get you out of the hobby shop and into a field you enjoy."

"Lawrence also applied."

"What in the world makes you think he is qualified?"

"Well, he does have a degree in geology. They are going to send a geologist to Phantasmagoria for sure."

"I forgot about that. I hope you are both selected." She giggled. "I would like to see the look on Todd's face when you both resign. He will probably inhale his mustache."

I laughed. "You are probably right."

25

Phantasmagoria

∞

"Debra, guess what I found out today!"

"I can't guess. What happened?"

"Spacex wants Lawrence and me to travel to The Kennedy Astronaut Training Center on Wednesday for a physical!"

"Congratulations Flynn, I have been praying that you would be selected."

I had just arrived home from work with a spring in my step and feeling like a kid whose dad had just been made the president of the United States. I fairly flew across the room, gave my new bride a kiss, and then dove for the phone to call Jerry. I told him the news with great pride.

"Congrats man! Does warthog know about this?" he asked.

"No, he doesn't. It has been tough keeping a secret from him for two weeks, but I guess he will find out when both Lawrence and I ask for leave on Wednesday."

"He will have a heart attack."

"No, he would never do such a service for others."

∞

"You are in great shape, especially considering that you work indoors."

"I do pushups every morning and I try to jog three times a week."

"If more Americans would do the same thing, then we would have more people to select from. Most of the people that have come in here the past three days have been extremely overweight and out of shape."

The doctor looked in my ears, eyes, nose and mouth.

"How many people have you passed so far?"

The Doc grinned. "Are you worried about your chances?'

"I uh, I just wanted to have an idea of what I am up against."

"Well son, I've given the okay on about eight dozen so far. If I give the okay, they go in for a small interview down the hall. If not, I show them the door and they are none the wiser till they get a phone call."

"What about me? Am I okay?"

"You are one of the healthiest people I have seen so far. Put on your shirt and go down to room 119."

"Why did they not tell anyone about an interview?"

"They do that so I can weed out the heavy and the sick. I guess they also want to see people in a more normal state of mind."

"Thanks Doctor."

"Don't worry, it's not a really difficult interview and remember, the longer they keep you in there, the better chance you have. Most guys are out in three minutes."

"Thanks again Doctor."

The doctor wrote something on my report, folded it and handed it to me telling me to give it to the interviewers when I got in there. I walked out into the hallway a little stunned from the shock of being told I had an interview with no opportunity to prepare. At least everyone was on the same level.

A receptionist was waiting outside of room 119 with a notebook. I handed her my report from the doctor. She read it than invited me into the room. It was a small meeting room with rows of chairs at one end and a long desk at the other. Seated behind this desk were ten straight-faced people. The receptionist gave my report to one of the people who quickly glanced at it before passing it on to the others who did the same. I sat down in the chair that the receptionist indicated and tried to conceal my nervousness. Once the last person had looked at my report the man who appeared to be in charge spoke up.

"Mr. Carson, your resume says that you are a pilot and you are licensed to give flight lessons."

"Yes sir, I have been flying for eleven years. I have flown everything from single-engine prop planes to twin-engine jets. I currently give lessons at the Hawthorne City Airport."

"Your resume also says that you work at Reeve's Hobby Shop. Can you tell us more about what you do there?"

"I work as a sales clerk and a model repair man. If have fixed airplanes, model rockets, helicopters, and submarines."

"Thank you Mr. Carson, that is all."

I shook hands with everyone, thanked them for their time, and left feeling slightly disappointed. The doctor said most people were interviewed for about three minutes, but I was barely in there for one. Out in my truck I called Debra and told her about what had happened. She also seemed disappointed, but relieved at the same time. I drove to the restaurant I had agreed to meet Lawrence at and waited for him to arrive. Fifteen minutes later, he entered the hamburger joint looking like he had just finished watching a ghost movie.

"Did the interview take you by surprise as well?"

He sat down across from me and picked up a menu.

"I about died when Doc told me to go to room 119. I wasn't interviewed but for a couple of minutes, but it seemed like eternity."

"They only asked me two questions and I was out so I don't think they will select me."

"I hope they do. I think that you are more qualified than me."

"I may be more qualified as a pilot, but perhaps they need a geologist more."

"All I know is that is we are not selected, we will never stop hearing about it from Todd."

"Come on Lawrence, now that you mentioned him I lost my appetite."

"I'm sorry."

∞

"Did our resident astronauts have a nice vacation yesterday? Where are your helmets oh brave spacemen?"

Todd had pestered us relentlessly since we had asked for a day off to pursue another career; not that working for Warthog could be considered a vocation.

"Be sure to say 'hello' to the Martians for me, will you?" Todd continued. "And why don't you guys leave early so you can go learn about anti-gravity and defense against aliens."

The phone rang.

"This is Reeve's Hobby Shop," I said. "How may I help you?"

Todd continued his ranting in the background. Lawrence studied an electric motor in his hand and tried to appear calm.

"Thank you, sir." I said. "I look forward to that."

I hung up the phone and started writing a note which I then put in my pocket. Todd looked at me with a furious curiosity.

"That was Bruce Chapman," I said.

"If he has something for us to repair, he can expect a two-week wait time. If he wants to buy something, he can find our store hours online."

"Mr. Chapman is the mission control supervisor for Spacex. Lawrence and I are going to Phantasmagoria!"

Almost as if we were on cue Lawrence and I pulled off our name tags and threw them at Todd's face. Lawrence gave a war-whoop and we charged for the door as the plastic badges slammed into Todd and caused him to reel back into a shelf sending thousands of electrical components everywhere.

"Bye you old warthog," I yelled over my shoulder as we exited the door.

Todd came to the door and yelled at us to stop as we drove off, but I never looked back.

Chapter 4

"Life-support system."

"Go."

"Communications."

"That's a go."

"Final Readiness is complete. Continue countdown at T minus thirteen minutes."

"Roger mission control."

I adjusted the flight plan that was attached to my suit leg. I looked around the tiny cabin at the other astronauts. I was sitting in the pilot's seat up on the operation deck along with three other astronauts. To my left was Ronald Duncan, the mission's captain. Ronnie was a middle-aged man with a firm jaw and the air of a leader. He had light brown hair and looked the part. He was the perfect man for the Captain's chair.

To his left was Clay Wilson, the navigator. Clay was twenty-nine and a bit of a geek. He had sandy blonde hair and wore round, black-rimmed, nerdy glasses. Clay was a really cool guy, but during training he had become known for his habit of reporting totally unnecessary information in addition to the useful information when asked for navigational readings. If we asked him for our position, he might continue until we knew our relative position to Neptune and Alpha Centuari. Despite this characteristic, Clay was the best navigator one could ask for.

Sitting on my right was Franklin Myers, our radioman. Frank was a retired army grenadier and had seen plenty of action in active service. He was also an amateur radio operator and electronics genius.

Matt Fletcher our geologist, Chris Carpenter our mechanic, and Hugh Burns the medic all sat below me on the systems level monitoring the screens that showed how our craft was functioning. My radio buzzed inside my helmet.

"T minus six minutes and counting," mission control said. "Begin terminal count autosequence."

At T minus six minutes we also had to align the flight computer. Clay and Chris started working together rapidly with mission control to align the gyros, accelerometers, and other devices used for navigation in space. A few seconds later they had everything set and began to check and recheck the calibrations and adjustments. One always wanted to know where he was in space.

At T minus five minutes the spacecraft was switched over to internal power. We went through another test sequence of our various systems to make sure they were working off of battery power correctly. I worked my controls and gave the go that they functioned properly as mission control called off each mechanic. At T minus three minutes I heard the liquid oxygen pumps stop and their nozzles move away from the three cores that made up the first stage of our Falcon Heavy rocket. A second later they moved away from our Raptor second stage. At T minus two minutes the Air

Force range patrol announced that we were 'go' for takeoff. The Spacex launch director also announced us 'go'. With one minute until launch, they flight computer was powered up and tested. At T minus fifty seconds the thrust vectoring controls were tested on the first stage engines. Then at forty seconds the propellant tanks on both the first and second stages were pressurized. Then the eternity until launch time began.

I fidgeted in my seat, just a little bit nervous. I fingered the flight plan and the flaps on my suit pockets. I looked around the cabin again. Everyone was fingering something, just trying to pass the time. That last thirty seconds felt like infinity. Finally, we heard the final ten second long countdown begin.

"Ten"

"Nine"

"Eight"

"Seven"

"Six"

"Five"

"Four"

"Begin engine ignition sequence"

"Two"

"One"

"Liftoff"

I felt a slight vibration at T minus three seconds and at liftoff I began to feel the faintest downward pull. The pull began to grow as the craft picked up speed. Launch control's voice crackled over the radio.

Phantasmagoria

"Mission control, liftoff is confirmed; we have liftoff; trajectory is nominal."

Ronnie fingered his mic button. "Crew condition nominal. Cabin condition nominal."

"Engine performance is nominal." Mission control added at the appropriate time.

I looked at the porthole in front of me. The sky slowly turned from blue to black as we left the atmosphere. After three minutes the first stage separated and the second sage shot us into a parking orbit around the Earth. I caught a quick glimpse of Phantasmagoria running away from us around the sun. There it was, with its year of three hundred two Earth-days, its days of twenty-five Earth hours, and its fifteen degree tilt to its path around our sun. Phantasmagoria had picked a path around the sun that was parallel to the other major planets. I could barely believe that I was actually going to visit the place. I had spent six months training, hoping, and praying that I would pass the final selection tests and be put on one of the first six flights to the alien planet. Everything had paid off as I was named the pilot of Phanta 1, the first flight. Lawrence had been placed on Phanta 2, which would launch six hours after us.

I listened to the various system checks and tested the equipment I was responsible for when my turn came. It took us only about thirty minutes to perform the check. As we came around Earth for the first time, we burned the second stage for a couple of minutes. This burn let us exceed escape velocity and begin to chase Phantasmagoria. There was lots of work to do, checking systems and trajectory, but finally things settled down and we could enjoy the trip. We took off our spacesuits and packed them into their respective lockers. I grabbed up a pack of freeze-dried Astro-food and began to dine. Some of the other guys were talking about what they would do when we landed.

"I am going to build a huge sand fort," Chris was saying. "I even brought a flag to fly above it." He pulled out a square of fabric for proof.

The New Frontier

"Are you crazy?" Hugh asked. "We will have tons of work to do. Most of your free time will be spent driving the rover around exploring."

"But if I make a fort, then I will have to dig. If I am digging, that will be useful for geology work. Am I right, Matt?"

Matt nodded his assent.

"See there? Even Matt approves of my fort!"

"Why don't ya'll do something productive?" Frank interrupted. "I am going to try to figure out how to make some form of aerial transportation using only natural resources."

"A fort would be productive." Chris retorted. "If Matt helps me, maybe we can make a composite out of the soil and form a shelter for storms."

"Both of those are very sensible ideas." Ronnie said. "I reckon a sort of contest will develop between the other camps to see who can create the most useful tools and machines using only resources from Phantasmagoria."

"If we work together as a team and use our combined talents, we will be able to make many things." I said. "We can build the fort/shelter first, then the airplanes."

"That sounds like a sensible plan." Ronnie agreed.

∞

"We are coming in about three degrees too high," Clay said. "Nose it down a little bit."

I tapped the stick down ever so slightly.

"That's better."

The cabin temperature began to rise as we entered the upper atmosphere of Phantasmagoria. The porthole was glowing with the intense heat. I studied the navigation screen in front of me closely. At their precise time the drogue chutes deployed and slowed our Dragon capsule with a jerk. A couple minutes later, I released the chutes and began to use the retro-thrusters to brake. I looked at the camera screens and chose a place to land. Actually, it was not that difficult since everything around was white sand. We hit the ground with a gentle bump and I shut off the thrusters. Everyone cheered.

"Houston, this is Phanta 1. We have made a successful landing." Ronnie said into his mike.

"Roger, Phanta 1. Have a great stay."

Ronnie turned to me and said, "Great job, Flynn."

I thanked him and began to take off my landing suit. We would have to wait an hour to do atmosphere tests and allow the Dragon capsule to cool off. Chris was busy running a system test of our spacecraft while Clay tested the magnetic field. After about fifteen minutes, the computer report confirmed that the atmosphere was safe to breathe. We patiently waited for the Dragon's skin temperature to fall to a safe level. Finally we got the okay from Spacex to proceed and exit the craft. Being the captain, Ronnie was the first out. He opened the hatch letting in a burst of cool, fresh air. He climbed down the ladder and jumped on the surface of Phantasmagoria. Since I was the pilot, I was the last to climb out of our faithful ship. Eventually my turn came and I stuck my head out into the bright sunshine. After being in a small, dark capsule for almost a month, my eyes had grown quite sensitive to light. I pulled my sunglasses over my eyes to try to dampen the blinding glare. Scientists on Earth had guessed that the sand would reflect light and cause something similar the snow blindness. It appeared then that they might be right. My eyes quickly adjusted to the light and I could handle it just fine.

I stomped my boots on the sand a couple of times as if to convince myself that I was not in a dream. The air was cool and a swift steady breeze blew from north to south. All I could see for miles were small, rolling dunes of sand. I bent down and picked up

The New Frontier

some of the sand. It felt just like Florida beach sand, except a little bit finer. A thick, puffy cumulus cloud drifted over us temporarily blocking out the sun. Several more giant clouds drifted across the sky on their everlasting journey over a barren landscape. The wind that blew them was steady and seemed full of life. Every breathe you took made you feel bigger, stronger, and smarter. Hugh said that the atmosphere of Phantasmagoria had a larger percentage of oxygen than Earth, as well as a slightly lower percentage of carbon dioxide.

I snapped out of my daze and began to help my comrades set up the scientific equipment we had brought with us. We set up a laser reflector similar to the one Apollo 11 astronauts set up on the moon. We also had an antenna tower for both audio and visual transmissions. Every couple of hours we had to check in with Earth and give them a report as to our condition. We also had a small solar powered rover that carried two people. A small excursion had been planned for the morning to try to find a local water source. The last things we brought with us that needed to be unloaded was a small nuclear fission generator and a couple of small tents. We set the tents up and anchored them to the ground with sandbags. Setting up the rest of the camp took up the remaining four hours of daylight, but finally the camp was in operable condition and we could rest.

I leaned against one of the legs of our Dragon landing capsule and ate a freeze-dried sandwich. I stared at the beautiful Phantasmagoria sunset which no man had ever seen before. Almost directly above me was Earth, looking like a bright dot about the size of a pencil eraser. I thought about the all of the events that had led me to this distant place and away from my lovely wife and dear home planet. It didn't start with the discovery of Phantasmagoria. No, it started when I was twelve years old and the world economic depression was in full swing. My father had been a business man and he had lost everything. He ended up working for the county, digging ditches to drain water from the roads. Unfortunately, this gave him a very negative view of life and capitalism. Of course, neither of these things was to blame, but telling oneself that does not put food on the table. I would often hear him come home from a hard day cursing America and fate. One day at the table I said that I wanted to be an engineer or a scientist and change the world like Albert Einstein or Isaac Newton. My dad had had a very difficult day and when he heard my plans, he practically hit the roof.

"You will never do anything out of the ordinary, neither you nor any other person! The times for revolutionary discoveries went with society as we know it. Look outside. The entire planet is in chaos. Everyone is either starving or homeless. We will never again have great artists or scientists! Stop thinking such things and spend your thoughts and energy on a way to put food on the table. That is all that matters now."

I will say that I never shared my father's pessimistic attitude. When he looked out at the world, he saw pain and suffering with no light at the end of the tunnel in sight. When I looked out, I saw opportunity for improvement. He saw what was, and I saw what could be. I remember going to bed that night thinking about his words and the inner pain they revealed. While he had intended to ease his suffering and 'blow off steam', he had actually helped my determination to change the world. I fell asleep dreaming of the good one person could do if they were just determined to finish at all costs. That determination stayed with me all through college as I studied mechanical engineering. Unfortunately, the costs of college had caused me to temporarily lose sight of my dreams for a couple of years while I worked a 'nine to five' job. Now I was sitting on a foreign planet enjoying a beautiful sunset and looking at a planet still suffering from the illness that had affected it for almost two decades. I looked at Earth feeling a sense of accomplishment while knowing that hundreds of thousands of people were looking across the gulf of space at Phantasmagoria with a feeling of hope. The uncharted planet had become the beacon at the top of the hill to the world, and now I was sitting on this beacon.

As the sun fell, so did the temperature. I went into my tent and put on some warmer clothes. The predictions were that the Phantasmagorian nights would drop down to a chilly forty-five degrees Fahrenheit while the days might exceed 120 degrees. While

The New Frontier

most people might not think much of forty degrees, I was a Florida boy and that is chilly. I could hear a couple of the guys talking in the tent next to mine. It seemed as though they were having a disagreement as to who should go on tomorrow's excursion. I chuckled knowing that the decision would be made by a dice roll, or so Ronnie had said. After a couple of minutes, Hugh and Chris crawled into the tent and curled up in their bags. I listened to them talk about our planet and its origin for a while, but I couldn't shake the thought of Debra. Here I was, on one planet that was free and peaceful and she was all alone on another planet that was suffering from economic disorder. Of course, much had improved in the year since the first scare Phantasmagoria gave us, but there were still many problems that needed to be solved. And I had left my wife all alone to face that cold, hard world. I knew that I would be returning in a few weeks and that she was getting my pay from Spacex, but I still felt a little bit guilty. Then again, what could I do about it? If I had not come along on this flight, I would still be working for Todd Reeves and making two-thirds of my current salary. And that did not include the bonus I would receive at the end of a successful mission. Yes, Debra would be just fine in my absence and I should try to get some sleep.

I guess I kind of dozed off, because the next thing I remember was Chris stirring me awake. It was still dark outside which was not what I expected.

"What is the matter?" I asked. "Did something go wrong?"

"No, it is just that Phanta 2 will be landing in ten minutes and everyone needs to be awake and ready in case there is an emergency."

I crawled out of the tent and stumbled over to the headquarters tent. Ronnie had a light on inside and I could hear the static of a radio. I almost collided with Hugh and Matt coming out of the tent.

Phantasmagoria

"Stay out here," Matt said. "We can see them enter the atmosphere. They will come in over there to the west and fly over us-"

"-and land thirty-two point five miles due east of us." I said reciting something that had been drilled into us from day one.

"Correct," he said. "It'll be here any minute."

We waited about three minutes, and then all of a sudden, there she was. It started out as a small speck of light, and then began to grow into a large flame as the craft streaked across the sky. When they were almost overhead the drogue chutes opened and began their work. Finally, as our buddies disappeared over the horizon, the chutes popped off and floated away on the desert breeze. A couple of minutes later we received a report that the landing had been a success and the Phanta 2 crew was waiting for their ship to cool before starting to set up camp. And that was how it was, every six hours a new crew would arrive on a different part of the planet and begin the tasks that each one had been assigned.

Chapter 5

    I awoke the next morning just as dawn was breaking on the horizon. I rushed out of my tent and stood gaping at a sunrise that no man had seen before. Everybody except Hugh was already outside viewing the spectacular sight. Matt woke Hugh up so that he could witness the once in a lifetime experience. There we stood; seven pioneers on a foreign planet, several million miles from our home planet watching our home star peek over the horizon of an alien planet. I must say, it was the most beautiful thing one could ever hope to see. The sun shone brightly through a high oxygen, pollutant-free atmosphere which painted a picture more brilliant than anything ever seen on Earth since Eden.

    We watched the sun rise until it was well above the horizon. Then we quickly set about our morning chores; some fixing breakfast, some rolling sleeping bags, and the remaining few preparing the rover for today's excursion. I was responsible for packing provisions for the rover team. I placed the small packs of food and water in their compartment then helped Chris check the electrical connections. We tilted the solar panels towards the rising sun so they could start warming up and charging the capacitors.

"The grub's ready." Frank shouted. Frank had been designated as our cook as the result of him losing a video gaming bet. Of course, our cooking only consisted of reconstituting dried food with water and warming it, but it still bothered him and he complained about it every meal. Chris and I hustled over to the solar heater and took our portion of the breakfast rations. On a sterile world we had to pack all of our food and ration it very carefully. Spacex launched supply capsules every week, but when they came we would have to share them with every camp on the planet.

As it was, we had enough food and water to last us a week on the surface and then a month back to Earth. In seven days they would launch the first of the return capsules. We would go into orbit in the Dragon capsule we had arrived in and dock with the return capsule which would be traveling slightly less than escape velocity. The return capsule would give an engine burn that would break us free from the gravity of Phantasmagoria and launch us back to Earth. Meanwhile, we had a ton of important work to do conducting experiments to learn about the planet.

After our small meal was over, the camp buzzed like an anthill with everyone scurrying around performing the tasks he had been assigned. Ronnie declared that Frank would accompany him on the rover scouting mission. Matt would stay in camp and perform tests on the soil. The rest of us would spend our time executing the thousands of things that NASA wanted us to do. I was in charge of launching several weather balloons to take atmospheric measurements. The weather balloons were made out of a black plastic which absorbed heat. The balloons would have worked better if they had been filled with helium, but helium would have had to be brought along and we simply did not have the room for such things.

The rover crew left a little after eight in the morning, Phantasmagoria time. We had brought clocks along that measured time in twenty-five hour days. After Ronnie and Frank departed, I quickly turned my attention to the balloons. As usual, there was a stiff breeze so the balloons headed south very fast once they were released. I sat down in front of my tablet terminal in the shade of the Dragon capsule and monitored the information the weather probes were sending back. I fed the information into the Dragon's main computer system which in turn relayed it to Earth. Since I was not trained in meteorology, most of the information meant nothing to me, but on Earth that was totally different. They could use the atmospheric properties and weather patterns to design aircraft which would be used by future colonies on the planet. Finally, about noon the batteries on the weather probes gave out and I put my tablet away.

We ate a quick lunch and returned to our studies. So far, two more teams had arrived on Phantasmagoria and the fifth team was due for entry in a couple of hours. Matt had confirmed that the sand was made up mostly of silica with large amounts of sodium chloride, a small amount of silver, and large amounts of a carbon crystal- a sort of soft diamond. He theorized that the sand would make excellent electronics if nothing else. He also said that about a foot down the sand showed large amounts of heavily decayed organic matter. This appeared to be a type of vegetable matter, similar to peat. He gathered samples of it to take back to Earth to be analyzed. He figured that if there really were organic nutrients in the soil, then the land might become arable and we could genetically modify some desert plants to thrive in the vast sands of our new planet. I thought that perhaps precipitation might help some of the nutrients in this sub-layer rise to the surface where grass and other plants could grow.

∞

Ronnie and Frank returned to camp about three o'clock that afternoon. They had confirmed that we were approximately six miles from the edge of an ocean. They then took water samples and spent the rest of their time exploring the coast. I looked over at Frank who was being very quiet and letting Ronnie do most of the talking. I could tell by the expression on his face that they had learned something, but were not sharing it. I decided I would follow up with that later. Meanwhile, we had the water samples to analyze.

The water appeared to have some bits of plant material floating in it. The organic nature of this material was confirmed by a microscope, but the cell structure was not defined enough for us to conclude that it was from a plant.

"Since this planet is from a totally different star system, it is entirely possible that the cellular structure of the vegetation differs completely from that which we would see on planet Earth," said Hugh, who was doing the analysis. "In fact, I cannot determine any sort of separate cells."

"I think the whole sample is just rotten," Frank said. He really had no clue about such things and was just trying to sound important. Everyone looked at him with totally confused expressions. He took a couple of steps back and said, "Okay, I'll just go work on... something."

Hugh stood up from the microscope and adjusted his glasses. "The sample is very decayed." Frank stopped and turned around with a grin. "Even though the material was frozen for much of its life, the process of entropy continues and being in existence for an unknown number of millennia has done it no good." He took the slide out of the microscope and inspected it against the light before continuing. "The only way to tell what type of plant it would have come from is to take it to Earth and have it analyzed in a real laboratory. It is too far decayed to grow a sample from what I have here, but they can copy the DNA and grow a clone."

This seemed to satisfy the curiosity of everyone else and they dispersed throughout the camp fantasizing about what type of life may have once roamed this strange planet. I decided to talk to Frank as soon as possible about their trip to the ocean.

I was not able to talk to Frank until after dark. I had decided to help him design the airplanes he wanted to build and used that to start a conversation. We were sitting out near the Dragon and using a solar/crank charged lantern for lighting. Frank had several pads of paper and was trying to identify what materials could be used to manufacture each piece. He figured that most of the plane's frame would be made out of ceramic which we could easily make from the sand. The wings could be covered in diffused-light solar panels which would provide more than enough power for the plane. I sat down next to him and jumped right into the project. My preliminary calculations showed that with our estimated wing surface we could power a whole community. The only problem to making aircraft was a lack of metals. We could extract silver from the sand which would help, but we needed copper, aluminum, and iron. I hoped that we would find these in the mountain regions to the north or on the ocean floor. We had been working for about a half hour when I decided to ask Frank about the day's events.

"Frank, I could tell by your face that Ronnie did not tell everything that happened today. What's up?"

His demeanor changed drastically when I brought the subject up. "You promise that you will not tell anyone?"

"I promise."

"Ronnie told me not to tell anyone, but if we don't tell the other camps will learn of it and they will tell."

"Tell what, for crying out loud?" I said, starting to get mildly annoyed.

"There were seashells and fish-like tissue on the beach."

That stunned me. I sat back and took it in for a moment. We had seen signs of small plant life and bacteria, but none of us had expected in our wildest dreams that Phantasmagoria had once been home to higher organisms. This confirmed that our new planet had once been in a happy orbit around another star and was cast away thousands of years ago, killing all of its life. If it had once been alive, then maybe we could resurrect it with plants and animals from Earth and return it to its former glory.

"Do you think there is a chance some of the creatures survived?"

"That would be very unlikely, but it is possible. Even if the surface was frozen, Matt says it would be possible for the very deep portions of the oceans to remain liquid from geothermal heat and motion. I suppose that plants and animals could survive for a while if there was something for them to eat under there. Nature has a way of protecting itself."

"Why would Ronnie not want to tell anyone about this?"

"He said it had to do with politics. There was something about the way people on Earth would receive it. He figured that we had better keep it hush-hush and the best way to do that is to not let anyone know. He would kill me if he found out that I told you."

"Well, you don't have to worry about him finding out. My lips are sealed."

"Thank you. I appreciate that."

We returned to our airplane project. I immersed myself in my work and tried not to think about the discovery on the beach. It was hard, but after a little while I was fully absorbed in the airplane. We drew out designs for the different parts and arranged the order that we would manufacture and assemble each part. Like I said earlier, the main difficulty we ran into was a lack of metals. Without metals, it would be next to impossible to make an electric motor. I figured that we would find iron ore once we dug deep enough, but Frank said that we should not depend on it and try to find a substitute. He suggested trying some sort of fiber optics, but I did not see how those would help us. After stewing on it for an hour, I bid him goodnight and went to my tent to try to catch a couple hours sleep.

∞

The New Frontier

I ate my breakfast the next morning in silence. Ronnie showed no visible signs of hiding something, but after working and training with the guy I knew he must be under a lot of pressure. Even a thing as simple as seashells on a beach could be of so much importance when those seashells were on an extraterrestrial planet. They represented life from another planet and another star system. If the news of those humble shells were to reach Earth, who knows what might happen. People might drop everything and demand that all available resources be used for further research. They could also demand that we abandon the planet and preserve its natural state therefore causing Earth's scientific community to lose valuable information. I could understand Ronnie not wanting anyone to know about the discovery and I could imagine that he was under orders from NASA and Spacex to keep silent if such representatives of former life were to arise. On the other hand, the world had a right to know. The public not just deserved to know of these findings, but the government had a responsibility to tell the public. However, I was an employee of Spacex and not of the United Stated government so I was under the rules, regulations, and responsibilities set forth by my employer. I mulled these thoughts over and over in my head while eating my tasteless breakfast. After we finished, Ronnie simply stated that today's only project was building a permanent building that could withstand strong winds and sand erosion. Reports from Earth-based telescopes said that large amounts of clouds were gathering on the poles and we should beware. Everyone received the task with delight at not having to maintain the schedule Spacex had given us and we set to work forthwith.

The first thing we did was find a relatively low and flat spot about a quarter-mile distant from our camp. Then we drilled down into the sand hoping to strike hard rock. We found rock about five feet from the surface. Ronnie took a rover trip to the ocean to fill up some water skins with salt water for the ceramic-making process. I helped rig the fission generator to release some of its extra heat. Finally about two o'clock in the afternoon our first batch of crude ceramic was formed. It was more like a tempered glass and somewhat brittle, but the tools we made were functional and we were quite proud of them.

Phantasmagoria

The first things made were shovels. We cast them in molds of wet sand and as soon as they cooled, four of us got to work digging out a hole in the ground. We dug straight down to the rock and then started digging outward so that our shelter would have a firm foundation. As we sat around the cooker that evening, we reflected on the day's events. Most of our hands were blistered and nothing had been done except for a couple of crude shovels and bricks. I looked around at my team and noticed the general negative attitude.

"Come on guys, cheer up," I said. "We may not have built much, but we learned a ton today."

"So we learned how to make bricks." Hugh said. "Woo hoo; we're as smart as cavemen."

"The only thing we may have made is bricks, but we learned how to form ceramic on a foreign planet with only the things we brought with us. The bricks we made today may not add up to much, but cities are built one brick at a time. Besides, we may be taking this from the wrong angle," I retorted. Frank looked up at me somewhat surprised.

"What do you mean the wrong angle?" he asked. "How else can we look at this?"

"I am referring to the storm shelter. We are trying to uncover an area of bare rock the size of our desired construction so that it has a stable foundation. We have only shovels to do that with. They would not have done that on Earth even fifty years ago; they would have used explosives. Since we do not have any explosives, we must change the construction of our shelter." I noticed that I had lost some of the guy's attention, but I continued anyway. "What if we were to dig four holes as big as the one we have now and use those holes as an anchor point for our shelter? We could fuse the ceramic bricks to the stone and therefor have the foundation without spending two years digging a hole." This got their attention and I could see my comrades mulling it over. I heard a lot of assenting and talk between the guys, so I continued. "And furthermore, why do we have to use bricks to build this thing? At our current rate of brick-making, it would take us months to build a shelter large enough to hold us seven without even thinking about the other teams. In that time we could have a pre-fabricated storm shelter sent to us from Earth. I propose that we use ceramic beams and plates instead of bricks. With our combined knowledge of structural engineering we should be able to come up with something that will work well." The guys seemed very excited about my ideas and we immediately sat down and started drawing new blueprints and diagrams of the project. We fell asleep that night with much optimism about the coming day.

∞

"Hey guys, I got some news for you." Ronnie got out of the rover and walked over to the work sight. We were almost done with the fourth hole in the sand. Two support beams had been raised and fused to the rock and a couple of guys were putting the horizontal frame beams across them. I got out of the fourth hole and walked over to Ronnie. By the look on his face I could tell that he was the bearer of bad news. We waited while the other guys finished what they were doing and gathered around. Finally we were all there and Ronnie began. "I just received word from Earth and to put it plainly, we have been marooned." He took a look around at the group of guys that he had led to this planet. I was quite puzzled as to his meaning.

"What do you mean, we've been marooned?" I asked.

"Basically, Spacex is not sending return capsules; they are sending some basic supplies to start a colony. The United Nations has implemented a colonization plan and is ordering Spacex to leave us here." He paused while that sank in. We had been double-crossed. Spacex had been tricked into sending us on a trip to a planet with no intentions of returning us.

"Do you mean that we are never going back to Earth?" Chris asked.

Phantasmagoria

"We may be able to get on a return trip in several years, but not in the foreseeable future." Ronnie said. I was infuriated. I looked at the rest of the guys. Chris looked heartbroken, Matt looked panicked, and the rest of the guys looked shell shocked. Ronnie continued in a take-charge voice. "In six months, Spacex will be sending groups of colonists and our families will have the first options of being on those flights. Meanwhile, we need to prepare for them. We have the opportunity to build a civilization. We cannot forget that in our emotional reactions to our present circumstances."

My anger left as I heard Ronnie put it into a different light. I nodded my head and said, "Ronnie is right. We have been done wrong, but it is giving us the opportunity of a lifetime. We cannot allow ourselves to be blinded by the injustice done to us by Earth. We are not just being given the chance to build a civilization, but to start over. We have the chance to throw away many of the problems of Earth. This goes further than building ceramic shelters and airplanes; we must form self-government. We must adopt a constitution and govern ourselves by it. We must protect ourselves by it."

Only Ronnie and I looked at the situation in any positive light right then. We were both furious at what our friends and employers had done to us, but we could both see past the immediate and into the future. I knew that over time the others would come to see it from our point of view. On one hand we had been abandoned on a planet with limited resources and a relatively harsh climate. On the other hand, we were the seeds of a civilization. We had been given a huge opportunity. While my goal had always been to change the world, I never thought I would be given the chance to start a new one. Unfortunately, the shock of possibly never seeing Earth again had left the rest of the crew in a rather unstable mental and emotional state. Ronnie and I would have to work together to keep their spirits up and keep them from giving up or going insane. The U.N and Spacex had done a good thing, but they went about it the wrong way. They should have told us from the beginning and asked for people who would not mind never seeing their home planet again. I supposed that their logic was that by not telling anyone, the applicants would be people like us who were scientists and engineers and for the most part had the drive to succeed, not pessimists who simply wanted to get away from everything and everyone. But no matter what their logic or intentions were, it would not shake the fact that they had still tricked us. I grabbed up my shovel and returned to work. Even though they had deceived us, we still had to make the best out of the situation. Giving up and starving when we ran out of rations was not the answer.

Chapter 6

"It looks great guys," Ronnie said. He jumped out of our storm shelter and strode over to me. "You did a great job keeping them in line," he said. It had been a difficult task to keep everyone motivated and out of depression after we got the news of being marooned on Phantasmagoria. I confess that I even started to suffer from some depression when I thought about everything I had wanted to do with Debra but now would never have the chance. I had promised to take her to the Grand Canyon, Niagara Falls, and Mount Rushmore. But she would like it here on Phantasmagoria. Sure, the landscape was plain and some things that we took for granted on Earth were much more difficult here, but we would adapt.

Our camp was not located near any fresh water so we had to distill all of our drinking water from the sea. The new shelter had a built-in still which would make the process more convenient than the setup we had used since our arrival. I looked up at the structure above me. It had the appearance of an alien spaceship. The living area was dome-shaped and raised five feet above the sand by four ceramic pylons. The bottom of the shelter was sloped in an attempt to reduce lift during high winds. The entire surface was also dimpled like a golf ball to reduce drag. Inside we had a water reservoir which contained 2000 gallons of distilled drinking water. The shelter could comfortably hold thirty people and in an emergency we could fit the entire Phantasmagoria crew of forty-two in it. It had taken us two-weeks to build the thing and we had completed it just in time as Earth satellites showed that the storms at the poles were moving towards the equator and therefore towards us.

I hopped on the rover with Ronnie and we headed back towards the Dragon. Matt was there and he was performing tests on some of the bedrock. So far it looked like it had high iron content and possibly contained gold. Of course, until the colonists arrived the only use of gold would be for electronics. We pulled up near Matt and got off of the rover.

"Have you found anything yet?" Ronnie asked. Matt had several test tubes with crushed rock and water in them.

"This rock is full of iron and bauxite clay, which is an aluminum ore. It also has seams of quartz throughout it and gold is intertwined in that quartz. I am surprised we are not finding more gold or iron in the sand as the sand is most likely formed from the crushed rock. I have also found traces of silver, copper, and magnesium."

"What will we need in order to smelt the metals and use them?"

"We have everything we need to smelt copper and iron to make steel, but aluminum is going to take quite a bit of stuff from Earth."

I followed Ronnie into the radio tent. He was going to talk to Spacex officials on Earth and I had nothing better to do so he said I could listen. He turned the radio on and gave the call sign.

Phantasmagoria

"This is Phanta 1 to Earth; this is Phanta 1 to Earth." We received a reply a moment later.

"Phanta 1, this is Spacex mission control, go ahead."

"I am going to give you all the camp report. The rations are nominal and we have finished the storm shelter project. We have approximately two thousand gallons of fresh water on hand and our still is working so we have a virtually unlimited supply. Our geologist has confirmed that the bedrock is high in iron oxide and we plan to start smelting it within the week. The rock also shows traces of copper and bauxite. We hope to obtain lime or something similar from the ocean rocks to make glass. I will report our progress as we undertake the solar project tomorrow. Over."

"That sounds great, Phanta 1. Reports show that the clouds around the poles are dissipating. However, what appears to be a sandstorm is approaching from the south. It is currently three hundred miles away and moving north at a rate of fifteen miles per hour. The average wind speed is estimated at sixty-five miles per hour. We recommend that you halt all major activities and move into your shelter. Can your shelter handle the entire Phantasmagoria crew?" Ronnie rolled his eyes, obviously disappointed at the delay the storm would bring in our projects.

"It could if they had a means to get here. Only three camps have rovers and there is no way that we could haul everyone here in twenty hours even if we had twenty-four hours of daylight. The rovers are solar-powered, you know. I am sorry, but they are going to have to tough it out in their Dragons unless they build something better."

"We understand, Phanta 1. Thank you for your report."

Ronnie turned the radio off and looked at me. "These guys forget what they sent us with. It is not like they gave us buses and roads."

The New Frontier

I shrugged. "What have the other camps been doing since they arrived?" Ronnie was the only one who ever knew what was going on at the other five camps.

"They have not done quite as much as we have. They have been spending most of their time exploring. Phanta 4 has a shelter built into the side of a mountain thingy they found, but they are the farthest away from anyone."

I looked at the map of the local region. He was right; Phanta 4 was way out there. We were a couple of miles away from the ocean, which was to the west. We were almost even with the top of the ocean. Five miles north of the ocean and fifteen miles northwest of us was Phanta 3. Phanta 2 was thirty miles east of us and Phanta 5 was ten miles northeast of them. Phanta 4 was forty-five miles north of Phanta 5 and Phanta 6 was seventeen miles south of Phanta 5. If the dust storm was going to hit the area, each camp would have to provide for themselves. I grabbed up the map and studied it closer. Each camp was within five miles of a water source, whether it was a lake or an ocean. I had heard that the lakes were salt water, but several fresh water wells had been dug. Perhaps they could make underwater shelters or something similar.

∞

I had put the map down and strolled out into the afternoon sunshine. For a moment I was transported to a Florida beach. I could see sunbathers, umbrellas, and sailboats. A jet formed a contrail across the sky and a slight breeze held a kid's kite aloft. Waves rushed up against the sand in an everlasting assault. I was jerked back into reality by Ronnie honking the horn on the rover. I climbed on our faithful transportation and we began the short ride to the shelter. I could hear Ronnie saying something about having the guys pack up the tents and move our operations to the shelter, but my mind was back on Earth. Oh, how I missed that planet and all of its imperfections. I have never been one to get homesick, but this was not just homesickness. I had been torn away from everything I knew and abandoned on a planet where I would have to work with

Phantasmagoria

my fellow men to produce the very things we needed to survive. There was no going to the supermarket; if we wanted it, we had to make it. I felt a resemblance to the early days of the American colonies, just after their separation from England. Just like them, we only had a limited society to fall back on. Thankfully, our advanced knowledge of science would keep us alive on this barren world. So far three camps had started farming with seeds sent to us on one of the supply shipments. One of the main advantages afforded us was that even though we were all men of science, everyone except maybe one or two had a strong practical side.

Ronnie disturbed my thoughts by nudging me with his elbow.

"Are you okay?" He asked.

"Yeah, I will be fine. Hey, how come you did not tell Earth about the gold that Matt said was in the rock?"

"Because Earth does not need to know that we have gold here." I noticed that he slowed the speed of the rover, presumably to give us more time to talk. He paused a moment before continuing. "I have been in contact with the leaders of the other camps and we all agree. Phanta 4 has put forth a constitution and it is waiting to be ratified by all of the camps. As soon as it reaches us, I will present it to the rest of the camp. It is very similar to the United States constitution, the only changes being that the bill of rights is more specific, plus a couple of changes that make it more applicable to our planet. We have also agreed to ratify a resolution which bans the export of our natural resources to Earth. We do not want Earth exploiting us for our wealth. They marooned us here and told us to start a colony, so we will make an independent colony. When they took the option to return to Earth from us, they took away their right to our resources, including the gold. The resolution states that we will ship them enough gold to pay for the equipment they sent us and that is all. This is Phantasmagoria, and we are now Phantasmagorians."

The New Frontier

I leaned back in the rover seat and thought about this for a moment. It sounded fair enough to me. Of course, the other option would be to hold the gold at a ransom and tell them that they could have the gold if they brought us home, but why? We would be the ones who did all of the work building the colony and fighting for survival. Right now we had no one to tell us what to do, no one to pay taxes to, we had plenty of gold and iron even if there was nowhere to spend it, and we could go as far as we wanted and never find another human being. We had the know-how to build anything we wanted, whether it be a palace or a jumbo jet. Why would we trade that for going back to Earth and all of its problems, just to live out the rest of our days at seventy grand a year? All of a sudden, I no longer felt homesick and I felt a little better about continuing the work ahead of us. We had a world to build and we might as well do it right.

∞

"We need to lift the wing up about six inches to get in over the brace," I said. We were installing the wing on PF-1, a prototype airplane. The PF stood for Phantasmagoria Flyer. The aircraft was about twenty feet long, with a twenty-five foot wingspan. The highest point on the plane was only six feet off the ground. It had an open cockpit which held four people in a precarious seating system that looked like a fish net with seats like the ones on baby swings. In front of the seat was an electric motor, which despite its primitive looks was powerful enough to drive a tractor and fast enough for a racecar. This motor was powered by solar panels on the wings during the day and 'capacitteries' (a combination of capacitors and batteries) would power it for up to three hours after dark in an emergency. The landing gear were skids made just for the Phantasmagorian terrain. It had taken us two weeks to get this far in the project and we were moments away from completing it. We had just finished making the wing and were installing in on the plane.

"That should do the trick," I said as the wing slid into place. Clay quickly jumped in with some nuts and put them on the bolt ends protruding from the wing braces. We attached the support

Phantasmagoria

wires to the fuselage and Hugh connected the wires between the panels and the motor. We tightened the aileron control wires and did a complete check of all controls and flight surfaces. I slid into the pilot's 'seat' and flipped the master electrical switch to the 'on' position. The other guys stood back, looking upon the tiny craft with pride. And why should they not? We had been on this planet for only five weeks and in two weeks of that time we had built a fully functional airplane out of only ceramics and glass made from the sand and metals smelted from the bedrock. The PF-1 was a beautiful example of scientific and engineering knowledge mixed with good ole' American ingenuity.

I slowly depressed the knob attached to the throttle relay. The propeller began to turn and then whirl with electric silence and speed. The only sound was the rush of wind and sand blown away by the propeller. Ronnie gave me the thumbs up signal and I pushed the knob in further. The slight hum of the electric motor slowly increased in pitch and the plane began to move forward. I went faster and faster as I allowed more electricity from the panels to the motor. After a couple of seconds, I felt the slight quiver of flight velocity. I pulled back on the video game style joystick on my right armrest and the aircraft rose sharply into the air. Once I had good elevation and airspeed, I turned around and flew over the camp and ceramic shelter. The guys waved enthusiastically at me and a couple ran along under me, but quickly gave up the chase since I was traveling in excess of one hundred miles per hour. I pulled into a circle pattern around our camp and landing site and climbed for altitude. I leveled off at about five thousand feet set a course due east for the Phanta 2 camp. By now I had increased my ground speed to almost three hundred miles per hour. The electric motor purred like a kitten with a full belly.

It only took a couple of minutes to fly to the Phanta 2 landing site. I nosed over into a dive to test its dive capabilities as well as loose altitude faster. My team had decided to keep the airplane project (affectionately called operation dragonfly) a secret from Earth and the other camps. The plan was for me to swoop down and surprise each of the camps in turn after I had flown the plane enough to verify its safety. I had enough confidence in its construction that the short circling maneuver and the high-speed trip to our neighboring camp was adequate for a test. I pulled out of the dive at around fifteen hundred feet and began circling the camp to lose altitude and select a landing spot. Someone had spotted the shiny airplane when I was around three thousand feet and spread the word to the others. When I leveled out the entire crew was outside their ceramic slab shelter looking at me while shielding their eyes from the fierce glare of the sun. I flew low over the camp once before landing. I had barely stopped the motor when the entire team surrounded the plane and began interrogating me curiously. I pulled my face mask off and climbed out of the plane. Once they recognized me most of their questions ceased.

"What the heck is this thing?" asked the team's captain.

"Well, it is a solar powered airplane we made over at Phanta 1 using only natural resources."

"This is fascinating," he said. "How long did it take you guys to build it?"

"About two weeks. It flies great and did not take too much effort to build. We could build them by the batch if necessary." The guys were looking over the plane with the awe of engineers who have found a new machine. I walked to the tail where Lawrence was examining the flight surfaces.

"Flynn, this is so amazing," he said. "How long will it be before you guys supply the rest of the camps with these things?"

"Ya'll are going to have to work for them," I replied, slyly. "We are not going to build these things without something in return."

"Maybe we will be able to trade some of the things we are working on for a couple of these." He never told me what they were working on, but it must have been pretty big if he mentioned trading them for multiple aircraft.

Phantasmagoria

I visited and talked with the guys there for a few minutes before taking off for the next camp. I made them promise not to radio any of the other camps about the plane so that I could surprise them. Flying was always a way for me to have peace and it helped my mind open up so that my thoughts could flow unrestricted and I could philosophize undistracted. This trip was no exception. Soaring high above the bare, desert-like landscape allowed me to think about everything that had happened since we had landed on Phantasmagoria. We had built our shelter just in time for the dust storm. The dust storm did not drastically change the landscape, although three feet of sand was mounded up around our Dragon capsule. After we had dug the capsule out, we did something that seemed to break our ties with Earth. We took the Dragon apart and used it to build a hydroponic greenhouse for vegetables. That was a very emotionally hard project for all of us, but necessary for our survival on this raw planet.

Once we were finished with the Dragon, we put all of our time and energy into the PF-1. The PF-1 was the not just a useful tool for our survival, but a symbol of our independence from Earth. It showed that we could make machines without help and resources from our home planet. Once we had built a fleet of airplanes, then Earth would realize just how serious and complete our independence was. They would panic in the same way England had when the United States broke away from the motherland. As soon as politicians learned of our mineral riches, we might have a fight on our hands. Of course, the Phanta mission crews and colonists would have the same advantages the colonial Americans had. We would know the terrain and how to survive. Even Ronnie had said that such a conflict was unavoidable and we should prepare so that we would be ready when occupy troops arrived in a year or two.

These thoughts kind of led me into a daydream about our survival on this planet. As Phantasmagoria was picking its path around the sun, the weather and landscape were starting to go through changes. First of all, the humidity was rising. In fact, many areas around the planet had received loads of precipitation. Matt said that once the sands and soils had some sort of vegetation

The New Frontier

planted, they might start developing topsoil that was farmable. I had heard that genetic engineers back on Earth were developing a grass suitable for Phantasmagorian sand. They were genetically modifying sea oats and giving them traits from other plants that would make them spread and grow faster. I looked over the vast beach/desert and imagined it covered in grass and trees. I could be happy here either way. The grass and trees would be nice, but I had started to grow fond of the sand stretching from horizon to horizon. Something about Phantasmagoria gave one an airy feeling; a feeling of freedom that no Earth scene could imitate. We must defend that freedom even at the expense of our lives. With the higher oxygen content in the atmosphere and water in the upper atmosphere, survival would be easier and if we were ever forced to bug out in the future due to an Earth invasion. We could live much happier and healthier here than on another planet.

As I flew to the camp of Phanta 6, I passed a large freshwater lake. I could envision a village on its bank with fishing boats rocking on its slight waves. I could almost see the rods and nets of the fishermen as they used their expert skills to feed their families. I could see a school and a church in the center of the town with children playing in the ceramic plate street. My dream was so realistic that I almost felt blinded by the sun reflecting off of the solar panels on houses and carriages. People strolled up and down the streets buying goods, talking, and simply being merry. I could see a couple of high school-aged youth kissing and carving their names in the bark of a tree on the edge of town. Around the outside of the village were several small farms with fields and livestock. In between two fields was an airstrip. I flew low over the flat space of land, circled around and moved into a position for the final approach. Of course there really was no ceramic tiled runway, but the sand was smooth and I landed the airplane with grace. I turned it off and got out. In my mind's eye I could still see the happy little village clearly. I walked a couple of hundred yards to the edge of the lake and strolled along the shoreline. A boy kicked his ball and it rolled past my feet. I picked it up before it rolled into the water and tossed it back to him. I kept walking towards the pier. A couple of small solar powered boats bobbed up and down, tied to the pylons that supported the pier. A medium built man of about thirty-five trolled his tiny boat up to the pier and tied it to a metal loop in the wood. He climbed out of the boat with a stringer of fish. I noticed that this man looked almost exactly like me except that he walked with the slightest limp. I waited at the base of the pier and watched as he passed me. I could have sworn that he was a twin brother of mine, but I had no brother. I followed him as he walked into the village with his catch. I nodded to some ladies as they passed me on their way to a meat vender. I almost lost the fisherman when he turned down a street, but I quickly caught up with him. He walked a couple of hundred paces to a small, but neat and tidy house with a well-kept lawn and flowers on the porch. As he ascended the stairs, a beautiful lady came and greeted him with a kiss at the door. The lady took the fish from the man and disappeared into the house. The man then left the porch and headed to a small garden beside the house to pick one of the plump watermelons growing there. I leaned against the fence bordering the fisherman's yard and watched him. I had noticed that his wife closely resembled Debra. In fact, she had even had Debra's beauty mark on her neck. I turned away from the house and began to retrace my steps through the cozy village.

Somehow this was no longer a dream because I could feel reality. This was more than a trance or hallucination. It felt real. It seemed real. I could tell that it was not real, but somehow it was. I left the town and returned to my plane which sat on the pavement near the runway. I climbed in it, scanned the sky for other planes, and taxied to the edge of the landing strip. I checked the sky again before letting the motor go all out. As I left the runway, I felt a cold chill of sorrow and loneliness fall over me like cold water. It seemed to cover my entire being and eat into my soul. I banked the plane over in the direction of the Phanta 6 landing site and opened the plane's throttle. I gave a quick look over my shoulder and saw the town still peacefully living there on the edge of the lake. It seemed so real. I finally shook my head and faced the front, determined not to look back again.

As I soared over the desert, I pondered what had just happened. I felt sad, but hopeful at the same time. I sorely missed

Debra. Thankfully she had decided to come along on the first shipment of colonists and I would see her in a couple of months. Maybe my desire to see her and hold her was causing me to see things that were not there. No, it had to be something deeper than that. The vision felt so real, no drug or emotion could cause it. I then thought about what had happened in the dream. Someone who looked like me only five or six years older had come in from fishing and gone to a house in which was a woman who looked like Debra, only five or six years older. Maybe my desires to build a successful society on Phantasmagoria had caused me to hallucinate as to what such a society might look like when it was fully built and functional. Even then I could not guess as to what the images of Debra and I meant.

By the time I arrived at the Phanta 6 camp I had been able to push aside the thoughts about my vision and return to my previous state of mind. I landed and went through the usual procedure of explaining the aircraft and promising to visit again. It seemed that each camp I stopped at was working of some project worthy of bartering for several aircraft, though none of them would tell me what they were working on. I could make no educated guess as to the honesty of their preliminary proposals, but each camp had shown significant progress in designs of shelter construction and ceramic making. The architecture varied greatly from camp to camp, but all shelters were fully functional and worthy of compliment.

From Phanta 6 I would have to fly to Phanta 3 which was my last stop, but the flight path would bring me over my home camp. I decided since I had two hours before sunset that I would stray a little to the north and fly over my lake again. It took me only a minute to reach it. I looked down at the still lake like it was an old friend. I did not see any village on its shoreline, though I felt its presence. I did not slow down but kept my speed and left the lake behind in a moment. As I zoomed over the Phanta 1 camp I dipped my wing in salute to my friend's waves and cheers. A moment later I was being congratulated and backslapped by the Phanta 3 crew.

Phantasmagoria

After I returned to my camp that night I refrained from mentioning my experience to anyone. I knew it would do me good to get it off of my chest, but the time just did not seem right. I ate my supper in deep thought about the future of Phantasmagoria. The camp seemed quiet; each man missing his family and longing for the day that the transport ships would enter our atmosphere. Everyone except Chris was married and a couple even had kids. I could not imagine what a kid must feel like being told that he is going to live on a new planet. I thought about the way that their classmates would both love them and hate them at the same time for the opportunity that they might never have. With the way children were, the young colonists would probably be bullied by the extremely jealous and treated like heroes by the only slightly jealous. Of course, adults were the same way, even though the bullying would be done in a more nonchalant way. I chuckled to myself when I thought about the way that adults never really stopped being childish, but simply changed the way they showed their immaturity. Perhaps it was not a matter of being childish or immature, but simply a matter of being human. As we aged, we showed our human identity differently. That was the only was someone could change so drastically when they turn into an adult, yet never change at all. I continued to turn these thoughts over in my head as I curled up in my warm bag in the shelter. I looked out the open ventilation window at the beautiful stars. Someday, one of those stars would fall to the planet and I would be with my lovely wife again. I looked down at a barren hill of sand that made up the horizon. This planet had a long way to go before it would be a world, but we were making progress by the day. Yes, we were advancing by leaps and bounds.

Chapter 7

  I climbed into the pilot's seat of the airplane and tested my controls. Everything worked fine. I flipped the main switch to the 'on' position and prepared for takeoff. I checked the runway and then depressed the throttle knob to takeoff speed and hung on for the ride. The PF-3 transporter sped down the runway and rose into the air, acting ever so slightly sluggish in comparison to its lighter counterparts, the PF1 and PF2. I looked back and checked the cargo rack. Sure enough, the two Dirt Demon rove-karts were riding smoothly. Once I had attained an altitude of five hundred feet, I banked slightly to the north towards Seeone.

  Seeone was the city built by the Phanta 2 and 5 teams for the colonists from Earth. The colonists were to arrive in a day and everyone was getting ready for their arrival. My job was to transport solar powered machines to the city from the Phanta 1 landing site where we manufactured them. I had airlifted twelve PF1 cruisers piggyback style to the town in the last week and was now hauling the go-kart-like cars we called Dirt Demons. I could only carry two per trip in the transport rack of my PF3.

The PF3 looked like the Wright Flyer with a rack in the fuselage for holding the cargo. On these planes, only the top wing had solar panels since the lower one was almost always shaded. We had made five PF3's and all of them but one were at the Seeone sight. It only took me a couple of minutes to fly to Seeone and circle for landing. Seeone was a small town consisting of about thirty ceramic buildings. Seeone was originally named Colony 1, but this was quickly shortened to C1, and then finally Seeone. Of the thirty buildings, five were large apartments, five were storehouses, five were greenhouses, five were 'factories' for processing different things, three were armored storm shelters, three were 'strongholds', one was a church, another was a school, and the last was a courthouse. There were also a couple of smaller constructions that were not buildings in the classic sense, but necessary for the survival of the colony.

I landed the plane and taxied over to the area where I was supposed to put the Dirt Demons. I got out of the plane and rolled the karts off of the rack and on to the ceramic plates that made up the tarmac. I pushed each one into a neat line beside the row of PF2 fighters and locked their brakes. It was extremely hot so I decided to walk over to the courthouse where we had a fresh water reservoir. I looked around the empty town trying to imagine what it would look like in a week when the new arrivals had landed and had settled into their new way of life. It would be interesting to watch a hundred people be suddenly transplanted on a new planet in a pre-built town with a government system already in place.

I got my drink of water and went outside to sit in the shade of the courthouse building for a few minutes before heading back to Camp 1 for more Dirt Demons. I leaned my back against the wall of the building and pushed my fingers into the dirt. In the previous months, large rain showers had sprouted ancient grass and weed seeds which were turning the sand into real soil. The rainwater already made the sand kind of stick together and act denser. The vast deserts and beaches of Phantasmagoria were making a change to pastures and fields without human intervention. Of course, the colonists from Earth would be bringing grasses and trees with them that would help speed up the process. I got up and started to walk back to my plane. All of a sudden, something dove out from behind the courthouse and jumped for me, yelling. I ducked and moved forward, grabbing the thing and flipping it over my head, judo style. I whirled around and maneuvered into a defensive position. Lawrence slowly rose to his feet, stunned.

"Lawrence, you almost scared me to death!"

"I didn't know you knew martial arts."

"I didn't know you were here."

"I was tending the greenhouses. I saw you sitting there and I could not resist jumping you."

"Well, in the future, just remember that it might not turn out the way you were expecting." He followed me out to the tarmac where I got into my plane and left him standing there to finish his chores.

∞

I watched as the red and orange parachutes opened above the plain landscape. The transport craft hung beneath the chutes, suspended like a spider from its web. It was a newly designed NASA transport ship, capable of carrying one hundred passengers and returning to Earth robotically. I could barely contain my emotions. I had not seen my wife in almost eight months, and there she was, a quarter mile away but I was unable to see her for another hour. The transport landed softly on the sand and released the parachutes. A couple of wild guys from Phanta 6 chased them down in Dirt Demons. The rest of the Phanta crews were gathered around one of the stronghold towers at the edge of Secone. The tower looked like the rook from a chess game and had been built as part of a defense system that would be fully installed at a later time. I was out on the tarmac in a PF3, ready to transport supplies from the newly arrived spacecraft.

The cool-down hour took eternity to pass, but finally we received permission to approach the craft. I accelerated down the runway and let the plane rise into the air as soon as I had enough speed to operate the controls. A couple of guys were driving Dirt Demons, racing to see who could be the first one there. The rest of the guys waited anxiously for their families and friends on the tarmac. In a flash I was landing beside the large NASA transport and rolling to a stop. I got out of the airplane as the gangway lowered and the first passengers disembarked. The new arrivals made a steady stream out onto the planet which they had never seen before. They spread out like tourists in a new park and began to examine their surroundings. I scanned the crowd expectantly for Debra, but she was nowhere to be seen. The guys in the Dirt Demons arrived and began talking to some of the colonists in macho tones. I headed into the crowd looking for my wife. Almost everyone was out of the transport except some of the crew and a pregnant lady who was being assisted by the Captain. I walked over towards the gangplank hoping to talk to some of the crew about her whereabouts and the cargo that I was supposed to haul to Seeone.

"Flynn!" said a sweet womanly voice which I recognized as belonging to Debra.

I looked around, "Where are you?"

"Up above you," she said.

I looked up and realized that the pregnant woman descending the gangway was my wife. I sprinted up the couple of steps to her side and helped her down the rest of the way. As soon as we were on Phantasmagorian soil, I wrapped her in my arms and kissed her for the first time since I left Earth. I felt as though I had completed my mission to this foreign planet. I had come here and helped make it possible for human colonists to come and live without each one performing all of the necessary steps of survival and now my wife was here. It was the happiest moment in my life except the time that

I slipped the ring onto her finger at the altar. I backed away a step and put my hand on her belly.

"I did not know you were…"

"I had the Spacex people keep it a secret. They almost held me back till a later flight because of the stress of spaceflight, but I told them I would come even if I had to stow away."

"You have no idea how much I have missed you." We began walking towards the transport plane which guys were stocking with supplies to be carried over to Seeone. "We made some additions to the shelter at Phanta 1 so you and the rest of the guy's wives can stay there with us."

"That will be nice. Oh my, did you guys build this plane yourselves?"

I nodded my affirmation and showed her some of the major characteristics of the plane. After it was fully loaded, we both climbed in and I flew it to Seeone where they unloaded it. I dropped Debra off there and repeated the process six more times until the transport was unloaded. I ate a quick supper at the transport around six, and then went to Seeone to help the new colonists move in. The plain ceramic buildings were received with mixed feelings by the colonists. Most of them thought that their new homes were fabulous and were very excited about the journey ahead of turning a plain, uninhabited planet into a world. Then there were some who thought the lack of architecture and culture was despicable. I really wished that these people had been left on Earth, but here they were. I just hoped that the struggle of survival would either convince them of their folly or they would grow sick of the place and go back to Earth.

We finished putting the supplies into their storage areas about ten o'clock that night. As I walked out onto the tarmac with Debra for our flight to Phanta 1, I looked back at the sleeping town. It was so amazing that just a couple of hours ago, the place had been uninhabited. It had the feeling of a ghost town. Then just like that, it was a real town, a village. It would take some time before

Phantasmagoria

everyone found their place and Seeone had the feeling of a functional society. They would have to develop a social structure and following the rules of human nature, a class system would develop. Thankfully, all of this growth would happen under the watchful eye of a very explicit constitution. The proposed constitution had been ratified and we regarded it as law. Like I said earlier, it was almost exactly like the constitution of the United States except that it put even more clearly defined restrictions on the government. I pondered the reactions that the people might have when they were informed of the constitution. I expected that delegates from Earth were on the flight, probably with their own socialist system to put in place. We had expected this and were prepared.

Debra did not say much as we flew to the camp site. It was late and she was very tired. Hey, I was tired too, but I had not been up since five with the anxiety of entering an atmosphere at thirty-five thousand miles per hour. We landed at our new home and walked to the shelter. The rest of the crew and their families had arrived earlier on ten-seater buggies. The main chamber lights were on, but the place was silent. I led my wife to our room where she fell asleep before I had turned out the light. I crawled in beside her and fell asleep thinking about the tasks we would have tomorrow.

∞

"We have strict orders from the United Nations. This is the government they put forth, and this is what will be installed here on Phantasmagoria." The speaker was a short, fat ambassador from Earth who had probably never even seen a picture of the planet before he came here. I was seated in the main chamber of the courthouse with the captains of the Phanta missions. Ronnie had brought me along to their meeting because of the morale support I had provided and because of my strong conservative convictions. We were in a heated debate with the representatives of the United Nations who had written down the guidelines for a socialist fantasy government, the kind that only exists in fictional writing. However, the Phanta crew members and an overwhelming majority of the colonists would defend our republic form of government even if it meant violence. Of course, we hoped to solve this peacefully, but there was no doubt that Earth would send troops if we did not succumb to their wishes.

"We have told you that the constitution that we presented to you has already been ratified and is the law of the land," Ronnie said in a stern, matter of fact voice. "We unanimously ratified it and ninety percent of the new colonists support it. There is no way that you all are going to shove that failed system down our throats."

"But who gives you scientists the right to decide what form of government this new world will take? After all, you guys are employees of Spacex and subjects of the American government and the United Nations."

I jumped up in a fury. "Correction, we are not employees or subjects of anything. We stopped being employees of Spacex when they failed to keep their part of the contract and return us to Earth. We used to be citizens, not subjects, of the United States but we gave that up when we were told that we were to be permanent residents of this planet. And we never were anything to the United Nations. This is Phantasmagoria, and we are representatives of the Republic of Phantasmagoria. And as to our authority, we have just as much authority as the American patriots did who overthrew the British. Don't tread on us!" The six captains clapped as I took my seat. The three ambassadors who had not yet spoken looked quite shaken, but their leader appeared not to hear a word I said.

"We represent the most powerful force in the known galaxy. If you dare to rebel against the provisions they have set forth, then I can guarantee that they will send troops to Phantasmagoria and each of you will be hung for high treason."

"Let them come," I said. "We are ready for them."

∞

Ronnie slapped me on the back. "You sure told them," he said. "I just hope that we can handle them when the troops arrive. The most important thing is to convince the populace of the magnitude of this conflict and how it will affect future generations. I believe one of the hardest things we will have to do is get all of the colonists on board for a fight. We are ready to defend a republic form of government with our lives, but are they?"

"We can convince them," I replied. We were walking on the outskirts of Seeone in the morning sunshine while watching the hustle and bustle of the new settlement. "I think that the first thing that needs to be done is controlling the transmissions to Earth. I know we stalled them for a day with the faked radio breakdown, but we need to keep Earth from finding out about the conflict for a couple of months. We have already proven that a few men can build a colony in a few weeks. We need to convince Earth of the exponential ability we have. Our rarest and most needed commodity in this conflict is people. Without people, we can't fight."

"So what are you suggesting that we do?"

"I think that we need to tell Earth that we can build colonies for a great number of people. We need to get them to send as many of those transports as fast as they can make them and we can return them. We need to cover the planet with as many colonies as we can before they find out about our rebellion. Of course, the first step is to control all of the radios 24/7. If Earth catches wind of this conspiracy, then they will snuff us out before can get started."

"You have really thought this plan through," Ronnie observed. "I really like the sound of it. Of course, while those colonists are making buildings, we need to be making planes and weapons. This is going to be a long, hard fight."

"But it will be worth it. What will we tell our kids when they are being taxed to death and living with no freedoms? The only thing we could say is that it was more comfortable to make transport planes and greenhouses. It will be a long haul, but I am up to the task."

"I feel the same way. I will have all of the radios except one disassembled and I will send a message to Earth telling them that we can handle colonists as fast as they can send them. After all, Seeone could hold three times as many people as we have right now without making any additions."

"I agree. We had better get a move on with our plans without any delay."

Ronnie set out to take care of the radio issue while I went to some find some of the Phanta 2 guys. I needed them to start teaching as many people as possible how to make the ceramic boards and to start making buildings. I would start training pilots and some of the guys could start teaching people how to make planes and buggies. I did not hope to finish all of these things in one day, but the sooner we started the higher chance we would have of being ready when the first troop transports arrived.

∞

Ronnie had been very successful in his venture. Earth was eager to reduce its population of ten billion even if the trips only took one hundred people each time. Ronnie told them about the capacity of Seeone and the rate of which we could build towns like it. NASA agreed to immediately open applications for a free ride to the planet and to send the colonists as fast as they could build the transports and fill them. The large transporters had been designed a decade ago for sending people to Mars, but when Spacex cancelled their mission, NASA did the same. They had five of these titanic ships in storage and could turn them out in a matter of days with the improved robotic manufacturing of the 2030's. Since the first transport had carried mainly family members of the Phanta crews, NASA already had a group of the general public trained and ready to go as soon as we said we could handle the population. NASA had already determined that it would take much less training in the future missions because the transports were completely robotic and accelerated at only two g's. On this first mission, they put the colonists through quite a bit of astronaut training as they thought it would be rougher on them.

The rest of the Phanta crew members had been equally successful in the training of their friends and relatives of the ways that we made our machines. Some of the guys estimated that we would have a new colony the size of Seeone ready in less than a week. The machines that would take the longest would be making planes and buggies. Those things were somewhat complex and took more time to manufacture than the flat ceramic sheets we used for walls and roofs. But nevertheless, we would have a full fleet by the time the next transport came in a month.

The longest thing it would take to produce would be food. Every transport would contain enough rations for its inhabitants to live on the planet for a month, but after that they would have to live off the land. The problem was that vegetables only grow so fast. The next transport would have a hydroponic meat incubator, but we still had to get many greenhouses in production. The Phanta crews had thought ahead and built the greenhouses first. The five in Seeone were already producing and would feed five hundred people, but many transports would be arriving in two months and it took six months for a greenhouse to produce. Right now we had a surplus, but if Earth sent colonists at the rate we wanted them to, we might begin to have a deficit when the greenhouses we would make in the coming week would have been growing for five months. Somehow we would have to buy a month.

I was sitting in the rec room in the Phanta 1 shelter, snuggling on one of the 'couches' with Debra. It was about nine o'clock at night and a couple of the guys were in there with their wives. Hugh was telling a ghost story.

"I tell you guys, the Ocala Medical Center is haunted. It is a very old building and I swear that the place has ghosts lurking around it."

"They are probably the ghosts of all of the people who died there. That is why I would rather die from a simple cut than go into a hospital," Matt said.

"Ghosts are nothing but figments of people's imaginations," Clay countered. "They have proven scientifically that ghosts are not real."

"But science can't determine whether or not there are spirits in a place," Hugh said, annoyed at the interruption. "And I know that there is at least one ghost there; the ghost of Hans Croft. Old Hans died there back in 1964 and he has haunted the place ever since. I saw him on three occasions." Clay snorted, but Hugh ignored him and continued. "The first time I saw him was when I was down in the basement one night. It had been storming all day and it was the perfect time for something like that to happen. I was down there just looking around because I was curious as to what the bottom of the building was like. I did not believe in ghosts until that night. I was down there poking around in the dark. I had a flashlight, but they have a very big basement so it did not light up a very large area. There were a lot of boxes and some old equipment down there, but nothing too interesting except the ghost of course. The place was very dusty so I was just getting ready to leave when I heard a noise coming from behind a stack of boxes. I shined my flashlight over there, but I did not see anything so I turned around and continued to leave. Then I heard the sound again. I turned back towards the boxes and started to approach them. Then I felt a slow draft of wind from just over to my left. I tell you, by then I was starting to get a little bit worried. I shined my light over to where the breeze was coming form and then I saw it, him I mean. He looked like just a white cloud floating through the air very slowly. If it had looked like a sheet I would had suspected foul play, but this was barely even a cloud or mist. It looked more like a heavy disturbance of the air, kind of like what you see above a hot barbeque grill on a warm day. By now I was starting to freak out, I mean I was scared. Then I heard a long, slow moan. The thing kept slowly coming towards me."

Phantasmagoria

"What did you do next?" Debra asked.

"I turned and ran as fast as I could. I did not even look behind me before I closed the door to the stairs, I just ran. I went out and got into my car and drove home as fast as I could. I was not about to wait around and get chased by that thing again. By the next morning, I had pretty much convinced myself that I was just seeing things and I went into work. Most of the others had heard about the legend of Hans Croft and a few superstitious people believed that I had actually seen him. I kept telling myself that ghosts were not real and that I had been wrong. After a week I had convinced myself of that and after a month I had almost completely forgotten about the incident. That is when the second sighting happened. It was very late at night and I was just getting in on my shift. I was walking around doing my rounds on the eighth floor when I saw the disturbance again. I was just strolling from room to room talking to patients and there he was, moaning and floating towards me down the hallway. At first I thought that it was a curtain blowing, but then I realized that there was no curtain there. I flattened myself against the wall and hoped that he would just pass by and leave me alone. You see, the legend says that Hans died because he was denied treatment there at the hospital because he had many outstanding bills that he never paid a cent on. As he was dying, he swore that he would continue to haunt the hospital and its staff for a thousand years if the world lasted that long. That is why I hoped he would just pass me by. After he died, people reported seeing a ghost around on and off for several years. Then the stories died down for a while. Then about every ten to twelve years people would start seeing him again. There would be about a dozen sightings over a period of about a year, and then they would all stop. I figured that it was just the legend causing people to see things until I saw him the first time. But to get back to my story, he went past me alright, but he went by only two feet away. I don't know what came over me, but as he was going past me I had a sudden impulse to reach out my hand through him. I did it. I was not surprised when my hand went through him, but then he stopped right in front of me and gave out a really long moan. I do not know how I did not pass out from fright

right then, but I managed to stay conscious. He paused in front of me for about ten seconds before continuing on down the hall.

"What about the third time you saw him?" I asked.

"The third time was not nearly as scary as the first two, but it was the first time two people saw him at the same time. I was taking a break with one of my doctor friends in the cafeteria when we saw him. This sighting happened six months after the second one. We were just pleasantly eating our snacks when we heard the moan. Sure enough, there was the disturbance in the air, hovering across the cafeteria. I guess that the reason I was not very scared was the fact that I was not the only one in the room. He went across the room and out the door. I have not seen him since, but several other people have reported sightings."

"So do you believe in ghosts now?" asked Clay.

"Well, I certainly do not argue against their existence. I guess you could say that that means I believe in them."

"Do ya'll want to hear about the time I was bit by a vampire at Niagara Falls?" asked Frank.

"Ahh, that is enough fairy tales for us two," I said, rising. I helped Debra up and we headed to our room to go to bed. We would have intense amounts of work to do in the coming days so we would require as much sleep as we could possibly get whenever we could get it.

Chapter 8

"We will have these three transports arriving in a week, then after that they will begin to arrive at a rate of one every three days." I was in a secret meeting with Ronnie and a couple of the other Phanta captains. We were discussing our future plans and the necessary steps that we must be taking to insure that the planet could support the rising population. Six more transports had arrived in the five weeks since the first transport arrived with our families, each carrying one hundred passengers. So far the food supply was holding out, but we were still expecting a deficit in about four months. Each of the transports was bringing a hydroponic meat incubator which could feed everyone if they ate nothing but meat, but after a couple of months people would crave anything plant made. However, it may be necessary for the people to live off of almost all meat while the greenhouses caught up with demand. Thankfully Phantasmagoria was very rich with minerals which in turn allowed us to make very healthy hydroponics. But even with the advanced hydroponic greenhouses, it still took time for the plants to grow and start to produce fruit.

Even with the possible food shortage we faced, there were many positive things to be thankful for. So far Earth had not the inkling of a clue about the rebellion and the conspiracy against the United Nations. We were building structures for new colonies at such a rate that officials on Earth assumed that we were all working together under their socialist system. Reality, however, was quite different. Certain people were starting to specialize in the different areas of manufacturing. Some smelted metals, some made ceramics, some cared for the greenhouses and meat incubators, some made solar panels, some made planes and buggies, and others made other miscellaneous things. While this would be expected in any society, a form of barter had developed between the different groups. Since most of the people would not end up using the surplus they made, the different groups began to exchange labor for metals, food and ceramics. They took the metals and ceramics and made them into buildings and machines. They then sold the finished products and traded them back to the smelters and ceramic makers in exchange for more resources. This cycle was very small and almost unnoticed, but it was there.

Another thing we had to be thankful for was the changes in the terrain. Phantasmagorian grasses and flowers had begun to grow in the valleys several months ago and now they covered the hills, turning the dirt from useless sand to soil. It would take several years before large scale farming could take place, but several small plots had been planted and were giving optimistic results. Some people traveling to the north and south had reported finding small plants that looked like tree seedlings. It would be years before trees were producing fruit, but it was encouraging to know that more than just grass had survived this planet's extraordinary inter-stellar journey.

Of course, it was still possible that even more than plants had survived the journey. By now the seashells were common knowledge and an item of curiosity. If plants had survived the frozen atmosphere, could sea creatures do the same? Many scientists thought that there was a very high chance of this happening. The chances of the oceans being completely frozen were very slim as the geothermal heat from the planets iron magma core would keep some of the water liquid even though the surface temperature was approaching that of absolute zero. When Phantasmagoria left its home star, the water would start to freeze first, then the atmosphere. When the oceans froze, they would start at the surface and begin to work its way down. This would then act as a corral and chase the fish and sea creatures deeper and deeper till either they were frozen in and killed or the water stopped freezing. If any creatures did survive, chances were it was only a couple of big ones who could survive the extreme pressures of the deep. These creatures would eat the smaller ones who got in their way as the pocket of liquid water got smaller and smaller. So far, no boats had been made so the only thing we knew about the oceans was from flying over it in airplanes and from pictures taken from satellites orbiting Earth. Someday we would make boats and solar sail over the waters, scanning them with sonar and infrared in hopes of revealing its mysteries. That however, would have to wait until things were completely settled out with Earth which might take several years.

Another possible threat that scientists feared might hinder our survival was bacteria. So far, the only bacteria that we had found were brought here by humans, but no one knew what might rise up. Each transport ship brought a large supply of anti-bacterial medicines and a couple of facilities had been set up to produce penicillin and other medicines in case something should break out and threaten the populace. With an increasing population, pandemic was a rising concern mainly as a result of our lack of hospitals and doctors. It was the rising population that sparked the need for our meeting today.

So far only about a hundred and fifty people out of the six hundred knew that an active rebellion was being performed against the United Nations. Everyone knew that the socialist agenda put forth by the Unites Nations was being largely ignored and most people did not care. Only a few spoke out against the capitalist turn that the new society was taking, but their hungry stomachs encouraged them to work and keep silent. Ronnie had done a good job of keeping everything a secret from Earth and I doubt that they suspected any conspiracy. Even if they did, it would take them just over a month to send any troops here and that would give us time to prepare.

We were busy analyzing everything that had been done and we were formulating a plan for the next six months. Right now, Phantasmagoria was just dropping around the sun away from Earth. This meant that the transports had to travel further to get here, but they were able to use the gravity of the sun as a slingshot to accelerate them towards Phantasmagoria. Because of this, NASA had built one hundred transport ships and would be sending them as soon as they could because once Phantasmagoria slipped too far around the sun, it would quickly begin to take the transports much longer to traverse the distance and they would start to push the amount of time that a transport could sustain its passengers. They would have to cut down the number of passengers on each flight and that would not be cost effective. NASA was sending the transports as fast as they could fill them up.

We did not have to worry about housing the arrivals or giving them planes and buggies, but food and weapons were a major concern. We would wing it on the food, knowing that we could always fall back on the meat incubators and live off of almost all meat if necessary, but the only advancements we had made in the area of weapons were on paper, and even those were very few. I proposed that we use the Birkeland–Eyde process to obtain nitric acid which we could then turn into potassium nitrate and use it as an explosive. The way the Birkeland-Eyde process worked was electricity was arced between two electrodes. This arc was stretched out sideways by a magnetic field to form a plasma disc in excess of three thousand degrees Fahrenheit. Air would then be blown through this disk. The arc would turn the nitrogen in the air into nitric acid. The nitric acid would then be dissolved in water and distilled to obtain potassium nitrate or saltpeter. Saltpeter was the main ingredient in gunpowder, but we could also use it in a more pure form for explosive shells. One of the major reasons that the Birkeland–Eyde process was not used very often on Earth was the fact that it used a ton of electricity to operate the machine. By the time solar sciences had improved enough for energy to be almost free, we had developed other explosives which were easier and cheaper to produce.

Phantasmagoria

I knew that when Earth began to send soldiers, they would be using very powerful explosives, laser guided shells, tracking missiles, and smart bombs. They would hunt us out with infrared and bomb us to death. But just like the British army was better trained and equipped than the American colonists, we could beat them by mobility and stratagem. We might have to resort to some guerilla tactics, but we would prevail. I refocused on the discussion in the room. It had been proposed that we use air weapons to launch explosive projectiles. Solar panels would provide the power to pressurize the tanks. After a little bit of debate, everyone agreed to the idea and we sat down to start drawing real blueprints. I was very excited. Now the things that had formerly been ideas were actually starting to become reality. The Republic of Phantasmagoria was really starting to take shape.

∞

Pheewmm!!! I released the trigger of the grenade launcher and lowered the device. A second later a large explosion on the hillside about a mile distance sent dirt and debris flying into the air. I rotated the cylinder which held the grenades then aligned the sights on the hill and fired. The grenade launcher I was using was air powered as were most of our ranged weapons. It held six high explosive grenades in a cylinder which had to be manually rotated after firing each shot. On my back I carried steel tanks full of compressed air. The air in the tanks would fire twelve shots before having to be replenished. I had a black solar cape draped over my back to provide electricity for the air pump under the air tanks. Air hoses ran up and over my left shoulder and across my chest to the launcher which rested on my right shoulder. The grenades themselves were about five inches in diameter and about ten inches long. They were shaped like a cigar and weighed about three pounds apiece. With the tanks, launcher, extra grenades, and sidearm my gear weighed about 75 pounds. It took a while to get used to carrying that much weight around while I ducked into holes and ran drill patterns, but it would be necessary for the inevitable conflict.

I walked back to the Dirt Demon where the other three guys of my squad where waiting. Each squad was made up of four people. One had a grenade launcher, one had a sniper rifle, one had a flamethrower, and the last guy had an anti-armor rifle. We each had a sidearm and an assault rifle in case of short range fighting, but we hoped that most combat would be done at sniper and rocket range. Lawrence was our crew's sniper. He was so good that once he detonated a grenade I fired at two thousand yards. The grenade was a dud and Lawrence saw it just sitting there so he took a poke at it and nailed it. Frank was our flamethrower. The flamethrower would not be used very often, but Frank still practiced faithfully and it showed. The anti-armor guy was named Jacob Barrett. He was almost as good a shot as Lawrence. His rifle was similar to the sniper rifle, only a lot bigger since it was designed to shoot an explosive charge through metal armor. I was the grenadier of my squad and one of the best grenade shots in the Republic's military. One of the things I liked doing was using the natural terrain as a weapon against the dummy soldiers and targets that we practiced on. I would try to cause landslides and rockslides to take out targets instead of simply relying on the grenade itself. As the grenadier, I was also the squad leader.

The tactic that our military seemed to rely on is the squads all working independent of each other. We ate, camped, and practiced as a squad. We did not spend much time working with the other squads and as a result we learned to rely on each other's weapons instead of someone else in another squad with the same weapons. There were no sniper crews, no machine gun nests or large strike forces, just the squads.

I climbed into the Dirt Demon and strapped my harness on. Lawrence was the driver at that time. He stomped on the accelerator pedal and we careened out over the grasslands. We had been on a 'practice tour' for a week now and we were ready to head back home. The type of Dirt Demon we were in was often called the 'War Devil'. They were painted an olive drab color and built a little bit lower to the ground. This was partially canceled out though by the fact that the tires were made slightly larger and had rougher knobs on them than the regular models. The War Devil had four seats, a storage trunk on the back for our supplies, and a six inch howitzer on top. The howitzer was for targets that my grenade launcher could not handle. The howitzer rounds looked a lot like my grenades, only they were several inches longer and packed a much bigger punch.

"That was some awesome shooting back there," Lawrence said as we silently zoomed over the terrain. The War Devil could go over one hundred miles per hour if needed, but we were only going about forty-five miles per hour then.

"Thank you," I said, taking a quick glance at the pressure gauge on my launcher. We were about two thousand miles to the south of the Phanta landing sites. The Republic had built arsenals all over the planet and we were responsible for making a routine check on one in our vicinity. I looked at the afternoon sun. The days were starting to get shorter and cooler. Since the planet was closer to the sun and not quite as tilted as Earth, we did not expect a very cold winter. But Phantasmagoria had surprised us more than once. I still remember the first time we saw grass growing on a hillside. We were driving a Dirt Demon around scouting for a good site to build Seeone on. Frank was the first one to see it. There was nothing really emotional about seeing the grass, but it helped make the planet feel more like our old home. I looked at the Earth over near the sun. It looked very much like Venus does from Earth, but a little bit bigger and with a moon. Even though I had been away from the planet for almost a year, I still felt homesick. I guess that I never would get over that. Even if Phantasmagoria were to grow forests, it would never have the majestic mountain ranges or the waterfalls of Earth. It would also never have the Home Planet feel to it. While we had discovered many things on Phantasmagoria that were similar to Earth, thankfully we had not found any volcanic activity or tectonic shifts. But if we did, what would we call them? We could not refer to them as earthquakes, and phantasmagoriaquake just would not cut it. I guessed we would come up with something if the situation arose.

The New Frontier

We approached the arsenal just as the sun was setting. The arsenal was nothing more than an underground dugout shelter. The door had grass planted on it so you would not know it was there unless you spent hours studying maps of the arsenal locations like we had. Lawrence stopped the War Devil and Frank and Jacob quickly jumped out, opened the door to the arsenal and made a quick check of its contents. Nothing was missing and no water had leaked into the underground cavity. They jumped back on to the buggy and Lawrence let it fly over the ground in a southerly direction towards the airfield that we would fly out of back to the Phanta area.

∞

"Welcome to Phantasmagoria," I said to the new colonists getting off of the transport ship. "We have buggies that will take you over to Hardinton where you can get a warm meal and some sleep." The colonists continued to stream past me while I repeated the message over and over. I wished that we had a public address system that we could tell everyone all in one shot, but we had not built anything of the sort yet and NASA did not bother to put one in their transport craft. These colonists had been selected from the New York area and I could tell that their New York attitudes would probably clash with the southern hospitality showed in Hardinton. But that would be for the citizens to figure out. After all of the passengers were off of the transport, I walked back to my PF1 and took off for Gabville. I do not know why they only wanted me to greet the passengers, but my orders said to head to Gabville as soon as I was done.

I did quite a bit of thinking once I was up in the air. Each of the towns on Phantasmagoria had elected a mayor and sheriff, but we had refrained from electing a president or congress. I did not understand it, but for some reason the Phanta captains had decided to put it off until the conflict with Earth was over. I did not know how they had managed to keep such a big secret as rebelling against the United Nations a secret, but no word had come from Earth on the subject.

Phantasmagoria

We had made huge advancements on the planet in our year here. The population was approximately eleven thousand people spread across twenty-two towns and villages. The transport that arrived today was the last until Phantasmagoria came around the other side of the sun and Earth could launch the ships backwards at us. Today's transport ship would have to wait until then to make its robotic journey back to Earth unlike the ships before it which usually began their trip within a week.

I landed the small plane on the strip just outside of Gabville. Gabville was one of the smallest towns and only had a population of about two hundred people. I walked down the street towards the courthouse where I was to meet the Phanta captains. Gabville was made up of almost one hundred buildings. People were now starting to build individual family houses instead of mass apartment style buildings like we first started out with. They were also starting to put fancy designs and colors on the ceramic instead of the plain block buildings that were first used. The vehicles were also starting to change. A slow, heavy dump truck passed me with a load of sand bound for one of the ceramic factories. Another dump truck followed it with red iron ore on its way to the smelting plant. We needed to elect a congress and form a currency. Right now everything was run off of a barter system which was working just fine, but we still needed a standard system. That would come with time. Right now our major concern was finalizing our independence from Earth. So far they had not come after any of our mineral riches, but perhaps that was because they underestimated our wealth. If they sent any military here, engineers would come and the secret would be revealed. As I continued down the street, I observed the quiet peacefulness that fell over the place. It seemed almost like the town in my vision. I would hate for this serenity to be spoiled by the noisy mining equipment that would be necessary for a large scale dig and the unscrupulous mine workers that would be sure to come too.

I reached the courthouse and opened the door. A couple of months ago someone had designed an air conditioning system and they were installed in many of the large buildings. Even in the Phantasmagorian winter it still reached ninety degrees in the middle of the day. Thankfully the planet was not as humid as Florida had been so most of the colonists could handle it just fine. The main room was brightly lit with electric lights. I walked to the back conference room where I had been told to go. Sure enough, there were the captains just sitting down to begin the meeting. I took a seat next to Ronnie.

"Gentlemen, the reason we are meeting is to discuss a dilemma that has arisen," began the captain of Phanta 4. "I have received intelligence that one of the United Nations ambassadors was able to get a hold of a radio and make contact with Earth." A gasp went around the room. "We knew it would happen sooner or later, but none of us wanted to see the day."

"How much did the ambassador tell them?" Ronnie asked.

"He told them everything. My source overheard the conversation and recorded it. I will not bother trying to play it back because the final results will be the same."

"How long ago was the message sent?" inquired the captain of Phanta 3.

"Unfortunately it was sent about three weeks ago. I had to pay my guy quite a bit of gold and copper for him to tell me what happened. He seems to be sympathetic to the United Nations. I was finally able to get him to talk when the price got high enough."

"That is terrific," Ronnie said sarcastically. "We have a whopping seven days to get ready for conflict."

"Look at the bright side," I said optimistically. "We have a large, very well-trained military and plenty of equipment. We also have the means of manufacturing ammunition almost on demand. We have a week to post scouts for their transport ships. We have been getting ready for conflict and now it is on its way. We can handle this. Did Earth give any indication as to their plans?"

Phantasmagoria

"Well, they said they had suspected something might be wrong and now that they had confirmation they would take action."

"So we definitely know that they will be sending troops. We also have a robotic transport ship here at our disposal. We can beef up its radar and post it in space to give us a little bit of a heads up when they arrive."

"Now you are talking," exclaimed the captain of Phanta 6. "There is no reason why we should look at this in a negative light. We have prepared and now it is time to make a blow for freedom."

We all gave a big 'hoorah' and the meeting broke up. On the way outside, Ronnie thanked me for my optimism and asked for a ride back to Phanta 1. We got on the plane and started the four hour long trip back to our home camp site. I was going to take a day off and take Debra to the Doolittle range. The Doolittle range was a huge canyon far to the north and was one and a half times as deep as the Grand Canyon. Since I would probably never see the Grand Canyon, this would have to do.

We landed on the permanent ceramic landing strip beside the Phanta 1 landing site. Our storm shelter was still ranked as the strongest building on the planet. The Dragon capsule which brought us here sat outside the shelter in a thin ceramic box to protect it from the elements. We got out of the airplane and approached the shelter. It had been a week since I had seen my family last and I was anxious for the reunion. I was just about to open the door when it burst open and Debra rushed out with Curtis in her arms. I embraced my wife firmly while Ronnie brushed past and went into the shelter.

"Flynn, I am so glad that you are here for a while. I have missed you so much!"

I took Curtis from her as we walked into the shelter. "I am sorry, but I will be leaving the day after tomorrow." The hurt and shocked expression on her face made me quickly explain. "One of the United Nations ambassadors got a hold of a radio. Troops from Earth should be here in about a week."

"I understand," she said. "When this is all over, can we just go somewhere by ourselves and you do something like making solar panels so that we can be together?"

"When this is all over, I will build us a small cottage on the edge of a lake and I will be a fisherman. There will be a small peaceful village there and the only worry we will have is which watermelon to serve along with dinner."

"That sounds absolutely wonderful," she exclaimed. "I think that is the way we were meant to live, not all hustle and bustle like so many people on Earth."

"That is so true," I agreed. "Now I have tomorrow off so we will fly to the Doolittle range like I promised you before I left. How does that sound?"

"I would love to do that. How will we travel with Curtis?"

"Frank's wife should be able to look after him. We will only be gone a couple of hours."

"Ok. I will make arrangements for that."

∞

"I had no idea that flying was so much fun!" Debra yelled above the rush of the wind. We had been in the air for almost an hour and she loved every minute of it.

"Now you understand why I do it so often and how I get so much thinking done." I replied. Debra raised her hands above her head. I looked over at her beautiful blonde hair streaming in the wind.

"How much longer till we land?" she asked.

"We should get there in another hour."

I was correct. In an hour exactly we saw the canyon up ahead. I dropped our altitude to only a couple of hundred feet above

Phantasmagoria

the ground. "Wait for it," I said. "Wait for it." All of a sudden, the ground dropped away from us and we soared over the mile and a half deep gorge in the planet's crust. I banked over and let the plane dive down into the canyon. We flew around from canyon to canyon for over an hour. Then I landed the craft a hundred feet from the rim and we got out to eat our picnic lunch. I carried the picnic basket to almost the edge and we sat down to eat. It was so peaceful there on the edge of that beautiful landscape. In one direction we had the canyon; miles and miles of rock that had been gouged out by billions of gallons of water from an ancient sea, and in the other direction there were hills, plains, and towns. Right then I would not have gone to Earth even if I had an instant teleport machine.

We finished our small meal and I packed up the picnic supplies. "Would you go back to Earth if you were given the opportunity?" Debra inquired.

I thought about my reply for a minute. That was a very difficult question and not one to be taken lightly. I looked around at the gorgeous landscape. "No, I would not," I said with final conviction. "Earth is good, but Phantasmagoria is a fresh start. We have the opportunity to work hard and see the immediate results. We are living a simple way of life that is almost completely unheard of on Earth. No, I want to stay here."

"Would you go to visit it?"

"Maybe, but it would be difficult not to bring back bits and pieces of Earth's culture and problems. Phantasmagoria is a place where we can let go of that and start over. If we were to visit Earth, it would be like pulling the scab off of an old wound."

Debra nodded. "I agree with you. That was one thing that I have been dealing with the past couple of months. Hearing you put your thoughts in that way makes me very happy that we can break away from Earth and not have to worry about giving up all that we have gained here." She had just finished speaking when I heard a sudden whoosh above us. We looked up to see a flaming object fly over our heads about half a mile in the air. It was headed in a westerly direction. It had just passed over our heads when drogue chutes opened arresting its flight. I looked at Debra in horror. This was a United Nation transport ship, probably the first one of several. The transport slowed down and started coasting towards the ground about five miles away from us. I ran over to the plane and grabbed up my spotting scope. Sure enough, the craft had United Nations painted on the side. The men inside did not even wait for the craft to cool down before they opened the hatch and began to exit. It was a regular NASA built transport ship and seemed to be inhabited by a full one hundred people. Debra arrived by my side.

"Is that the United Nations?"

"Yes it is. We need to get back to Seeone as fast as possible. Chances are they have ships all over the planet and will be set up to make a coordinated attack by nightfall. We will need every minute we can get." I gave the ship one last look through my spotting scope. The men were unloading munitions and field pieces. Debra and I got into the plane and we headed towards the settlements to spread the news like Paul Revere had done so many years before us.

Chapter 9

Debra and I were soaring through the air at almost three hundred miles per hour. On the way back to Seeone we passed two other United Nation transports. Since we were high and going very fast in a small white aircraft, none of them saw us. I kept pushing on the accelerator knob hoping to squeeze any extra speed out of the plane. It only took us about thirty minutes to get to Seeone. The plane had barely stopped moving before I was out of it and running. I ran through the streets to the courthouse yelling at the top of my lungs that the United Nations had arrived. The entire populace of Seeone knew what that meant so people started running from house to house getting their neighbors. I sprinted up the courthouse steps and burst through the door. Ronnie was standing there talking with the mayor.

"Flynn," he exclaimed. "What the heck is your hurry?"

"The United Nations has arrived. I saw one of their transports land near the Doolittle Range. I passed two more on my way here."

The New Frontier

Obviously our sources had mistaken when the message was sent to Earth. Or they had lied, being sympathetic to the U.N. as we suspected. Ronnie raced for the door yelling instructions to me over his shoulder. I sprinted behind him as he ran for the airfield. I was to go from town to town and alert them of the danger while he gathered the Phanta captains. We had agreed that when the United Nations arrived all of our operations would take place from Seeone since it was in a central location from the Phanta landing sites and since everyone who lived there was on our side. Debra was standing by the PF1 when Ronnie and I arrived. I told her to ride with Ronnie back to the Phanta 1 camp while I went on my Paul Revere like mission.

For the next three hours I flew from town to town spreading the news and dispatching scouts to try to locate other United Nations landing sites. Most of the scouts strapped on oxygen bottles before taking off so that they could fly at very high altitudes and therefore travel undetected as well as see further and cover more ground quicker. Finally, I had warned all of the towns. I flew back to Seeone where the Phanta captains would be holding a conference as to decide our next move. I needed to get to Seeone and help as quick as possible since Ronnie had ordered strict radio silence and all news and information would have to be carried by courier aircraft.

I entered the conference room and took my seat beside Ronnie. Everyone was there and they were in a heated debate. Half of the room wanted to attack the landing sites we already knew about, and the other half wanted to wait until the entire planet had been scouted and we knew were everyone was at. I broke the argument with the suggestion of parlay.

"Why don't we go to the closest landing site and attempt a parlay? Chances are they already have radar set up and scouts placed so an offensive would be detected. We know that they came knowing that we are very heavily armed. We should try to talk to them first and maybe we can arrange a deal with them. At the very least we can get an up close look at their weaponry and so we will have an idea of what we are up against."

Phantasmagoria

    Everyone liked my idea of parlay so we decided to attempt it. Ronnie and I boarded my PF1 along with the captains of the Phanta 3 and 4 flights and we headed off in the direction of the closest United Nations transport that I had seen. We flew about three hundred feet off the ground in hopes that they would see our somewhat primitive looking airplane and underestimate its potential. While pictures of the PF1 had been sent to Earth, we had kept its capabilities a secret. Of course it was possible that the information had been leaked out at some point, but it was worth the try. Ronnie was using my spotting scope to survey the land around us as we headed towards our enemy. Before long we could see them in the distance. They had finished unloading their transport and were setting up their artillery in batteries. I could tell that the transport ships had been made with smaller fuel tanks to give more space for cargo. This led me to believe that they were not planning on returning to Earth, at least not with those transports. Terrific, we had not just an army to deal with, but an occupational force.

    I set the plane down softly on the soil and taxied over to the unwelcome invaders. We had the foresight to attach white flags on the wings as a symbol of truce. We jumped out of the plane and advanced towards the camp under a similar flag. Ronnie suggested that we stop just a little ways outside of their camp and let them send a party to meet us. Sure enough, after a couple of minutes four men in officers uniforms started walking towards us carrying a white T-shirt tied to a metal pole. When they got close to us, Ronnie was the first to speak.

    "Hello, and welcome to Phantasmagoria," he said.

    "Thank you," replied the commanding officer. "But I assume that you know why we are here."

    "You are correct; we are fully aware of your intentions and advise you to consider doing otherwise as our force is armed and ready to defend our liberty."

    "I assure you that that is impossible. While we do not have anything personally against you, we have orders to put down the rebellion and establish the government provided by the United Nations."

"We are ready to die to defend our republic form of government. Just like the founding fathers of America, we will confront the strongest military power of mankind for our freedom. I just have to warn you, commander, do not underestimate our strength."

"You are definitely a determined people. You warned me against underestimating you. Well, at the same time do not underestimate us. Just because the American colonies were able to overthrow Britain, do not assume that you will be equally successful. Times have changed and we are well-equipped."

"We are not assuming anything. We know that this will be a long and bloody conflict and that is why I was hoping that you would consider letting us alone and saving lives on both sides."

"I do not have the authority to cancel our mission and even if I did, I would not. Shall we consider our parlay to be finished?"

We walked back to our plane feeling just a little bit sad. This parlay would mark the start of the first war between different planets. Even though we had all been born on Earth, by moving to Phantasmagoria one gave up that heritage. And now that we were officially at war with Earth the final ties had been broken. We were now Phantasmagorians and had to live as such. I looked back at the United Nations transport and the artillery batteries they had set up. They had radar on top of the transport scanning the sky for hostile aircraft. These guys definitely knew what they were doing. I was not very knowledgeable on military sciences, but I knew enough from practicing with our weapons that the guy attacking such a position would have a fight on his hands. Chances are most of those artillery shells were laser guided so all they had to do was have some guy hiding out in the grass and they could shell something miles away. I looked at the grass as we trudged to the plane. In some areas it was two feet tall. Someone in a ghillie suit could get mighty close to their camp without being seen. That fellow could probably carry and plant a larger load of explosives in one of those batteries than we could lob over the hills with the air powered artillery we had made. He would also probably have a larger chance of getting away from the camp and surviving the attack than if we shelled the camp with a bunker of artillery.

Of course we could always bomb them from the air. We had had a difficult time making an accurate bomb sight but some of the guys had good success dive bombing. I just hoped that the United Nations did not bring any aircraft with them. A couple of B-5 Stealth bombers could eradicate a whole region, if not a continent. We would need to set up aerial surveillance and if they started bringing out planes, then we needed to make that landing site a priority target.

We climbed into the plane without a word and took off back for Seeone. I knew that it would be wiser for most of the military to abandon the towns and head off into the wilderness where we would be less detectable. When we landed at Seeone the rest of the Phanta captains were waiting on the tarmac for us. We told them the news about the United Nations refusing to give up the fight. They then informed us that almost one hundred landing sites had been located all over the planet. That meant they had one hundred heavily fortified artillery camps to twenty-two unarmored towns. It was not very even, but what else could we do?

We went into the courthouse to hold another meeting. We had to make a fast plan then quickly put in into action before the United Nations could attack. We wanted the first move. "We need to strike out against them before they can launch an attack on us," Ronnie said, starting the meeting. "They have quickly set up batteries and defenses, but at the same time they have probably all been awake for many hours and are very tired after all of the work they have had to do. Even though they have those batteries in place, they are still very vulnerable. Each of those camps is in a very condensed position. A couple of well-aimed shots from field howitzers would snuff them out. They also do not have very much armored protection. We can assume that within a couple of days they will have thick ceramic shields protecting their artillery and other weapons. What do you all think?"

The captain of Phanta 2 stood up. "While you guys were gone, we received reports from the multiple scouts that we dispatched over the planet. Like we told you earlier, they located almost one hundred transport sites. These transports seem to gather in groups of between five and ten with several miles in between them. These clusters are generally separated by fifty or sixty miles. We know that their artillery is capable of sending a projectile that far. The conclusion is that each landing site is protected by its immediate neighbors and also by the sites in the clusters nearby. In chess it would be like attacking a zigzagged row of pawns. If you attack one, then the others are right in range to return the favor."

"In that case," I said standing up. "We need drive by shootings. Their artillery would have a hard time shooting at a moving target that is only several hundred yards distant. They would resort to small arms, while a couple of lobbed shots from our War Devil howitzers would send them up in smoke."

The guys liked my suggestion. "They would quickly catch on to our strategy and set up better defenses," Ronnie warned.

"I know that," I replied. "But is would serve to eliminate the sites nearest our towns. We need to act fast. We need to dispatch couriers to the other towns and tell them to strike only the nearest camps. Tell them to have the squad's sniper drive, the flamethrower operate the howitzer, and the grenadier and anti-armor guy throw small-arms shots at them as they go past. You should assign ten squads to each camp so that there is plenty of fire-power."

"You want ten squads per camp?" Ronnie exclaimed. "There are barely ten squads per town!"

"That is my point," I said, calmly. "They will only have the opportunity to hit one or two camps before the United Nations catch on to our plan. We might as well send them all to one camp at a time and therefore guarantee that the camps we attack are eradicated."

"I see," he said. "Let's do it."

∞

I held my grenade launcher close as we sailed over the small hills and ridges that covered Phantasmagoria. We were leading the pack that would strike the United Nations landing site we had parlayed with earlier. I looked back at the other nine War Devils. The straight expressions on the soldier's faces told me that they were determined to make the United Nations regret ever setting foot on Phantasmagorian soil. Lawrence led us between two low hills in a path that would put us right next to the camp. All of a sudden, we shot out passed the hills and into full open view of the target. Frank was on top of the War Devil and tucked down under the field howitzer in the seat meant for its operator. As soon as we were in sight of the camp, he let go with the first shell. I could tell that we had taken the soldiers by surprise. The howitzer shell shot right into the midst of their artillery, sending men and weapons flying. I fingered the trigger of my grenade launcher and took aim at the transport. We had approached the camp from the south, so I waited until we were on the southwest side of the camp. I held my fire until that split second opening where parts of the transport's fuel tanks were visible. Then I fired. The grenade streaked out and struck the fuel tanks, a bull's-eye. The fuel tanks were very low on fuel, but there was just enough to do what I wanted. The liquid hydrogen ignited with a flash. I saw the shock wave ripple across the grass with the fireball following it closely. The explosion covered the camp before rising up in a small mushroom cloud. The drivers slowed our War Devils to a stop in pure amazement and shock as the black smoke cloud rose into the air.

"I never even thought about hitting the fuel tanks," Frank said.

"Never mind," I shouted. "It was a lucky shot. The other camps will be ready for us in a few minutes. We need to move!"

Lawrence floored the accelerator pedal and we began the ten mile trip to the next camp. I reloaded my grenade launcher. The last thing I wanted was for everyone thinking that the transports could only take one shot and therefore ignoring the artillery batteries. There was only a very narrow opening on the sides of the ships were the fuel tanks were visible and if everyone spent all of their time aiming for that spot, then the hostiles would eliminate us with machine guns. About fifteen minutes later, the second camp came into view. Just as I had predicted, they were ready for us and quickly opened fire with M-16's. We were approaching the camp from the southeast across a relatively flat area. We were out of accurate range of their small arms, but they were within perfect range of our howitzers and grenade launchers. Lawrence stopped the War Devil and the others pulled up beside us about twenty feet between each one and we began our onslaught. I kneeled down beside the right fender of the vehicle and opened fire, trying to lob my shots beside the transport which was in the center of the camp and soften up the other batteries for when we moved around to hit them. I could hear the rhythmic whoosh and pop sound of our air weapons as we pounded dozens of rounds on them. The snipers looked through the smoke and tried to pick off the figures that ran around trying to defend themselves. The anti-armor guys shot at individual guns to disable them. It took us only a few seconds to completely decimate the batteries on the side of the camp that faced us. We jumped back into our War devils and hustled across the terrain and around the camp to attack another side. I shoved more grenades into the cylinder on my weapon while waiting for a clear shot. An RPG streamed out of the camp and exploded in front of our War Devil flipping the front end up in the air. The small vehicle soared over the crater in the dirt and would have flipped over backwards if the back wheels had not struck the ground on the far side of the crater, righting us. Lawrence stomped the pedal sending dirt and grass flying behind us and we shot forward, the War Devil seeming undamaged from the shock. I pointed my grenade launcher in the general direction of the hostile grenadier and squeezed off the shot. I replaced a round in the empty chamber as we rounded the corner and came into view of the rest of the camp. A good portion

Phantasmagoria

of the artillery guns on this side had already been disabled by our explosives, but there was still quite a bit to shoot at. I could see men on top of the toppled guns with automatic rifles, hoping to destroy the foe which threatened their lives.

I looked at these desperate men with a feeling of sorrow. These were men just like us who only wanted to survive. Most of them had probably joined the armed forces with a desire to help humanity and wanted nothing more than to perform drills and get a paycheck. Now they were here on a foreign planet shooting and getting shot at by a group of people who they probably felt sympathetic towards. Nevertheless, this was war and it was either kill or be killed. A bullet ricocheted off of the front bar on our War Devil, chipping it. I settled the grenade launcher on my shoulder and fired four rounds at the men on the wrecked guns as fast as I could rotate the cylinder and squeeze the trigger. The grenades hit their target and the machine gun fire stopped. Lawrence spun the wheel and we headed straight towards the smoking camp. I slid out from under the grenade launcher and grabbed up my combat rifle. Frank let another shell fly from the howitzer to stun them as we approached. The War Devil hit the dirt embankment they had set up in front of the artillery and went up it like a ramp. We soared over the crushed weapons and landed within their camp, testing the buggies' suspension to its max. We grabbed rifles, jumped out of our vehicle, and charged the bottom of the transport to finish off any survivors of our initial and secondary assaults. Two other War Devil squads made the jump and joined us as the rest of the guys parked outside to come in on foot.

In the heat of battle, you really do not think, you just act. I charged the transport staircase, spraying a burst of machine gun fire at anything that moved. I approached the transport's doorway, cut the pie, and moved down the hallway towards the main cabin. It only took us five minutes to finish off the remaining soldiers. None begged for quarter, nor did we give any. They all went down with a smoking gun in their hands, but fortunately we were warmed up for battle and took no casualties. We gathered outside the transport to regroup and discuss our next move.

"They will be prepared for us at the next camp," I said wisely. "We should pile up all of the ammo and whatever equipment was not destroyed and thermite it. Then we should get the heck out of here." The rest of the guys agreed with me so we set about gathering the ammo that had not been destroyed. We could use some of it in the weapons that we captured, but the rest we piled up to destroy. We did find quite a few claymore mines which would be of great use to us so we kept those. We also gathered the all of the M16's and handguns and as many magazines for those as we could find. All of the artillery was beyond repair so we left it alone. We got the three War Devils out of camp and the last guy out set a couple of thermite canisters to detonate with a time delay. Then we headed back to Seeone before the ammo began to cook off and send projectiles everywhere.

∞

"The first strike of the Republic was a smashing success, pun intended," the Phanta 5 captain said.

"We kicked their tails!" exclaimed the Phanta 6 leader. "We showed them alright!"

"We can expect retaliation," I said calmly. "Remember, we only eliminated twenty-five percent of their forces. They will likely shell several of our towns."

"What do you suggest we do about it?" inquired Ronnie.

"We should have planes in the air at all times ready to dive bomb any camp that shells us."

"That will be almost impossible during the night hours," Ronnie said. I looked outside the window. We only had about an hour of daylight left.

"There is no time to waste," I said. "They might shell us or they might make a ground attack. Whatever the case is, we need to evacuate our planes and weapons to the battery points we planned outside of the towns. We cannot have them destroying our forces in one swipe. That is exactly what will happen if we have planes outside on the tarmac."

"That is a smart idea," said one of the other guys. "We have got to get out of dodge or this will be the shortest fight since the six day's war. The probably are building a permanent base on the other side of the planet and are ready for the long haul. We need to move our military forces into the wilderness and settle down for a fight. A long, hard, desperate fight for freedom."

"We had better hurry," I said. "We have only an hour of daylight left."

Chapter 10

"We have received a report that United Nations forces are now occupying Gabville," said the Phanta 4 captain. "They rolled into the town last night and forced the small resistance there to retreat. Four Phantasmagorians were killed and two more were wounded." He paused and looked around the room before continuing. "Gentlemen, there is no reason why we cannot take Gabville back. It is within easy striking range and the troops are ready. We should not stand by out here in the wilderness and let them take over our civilians." He was right. The United Nations had been here for a week and since the day of their arrival we had not performed any real missions. A couple of guys had snuck into one of their camps and planted some explosives and another group of guys had shelled an armored column with great success, but that was it. And during that time, the United Nations had overrun the opposition and occupied four villages. I felt as though the Republic was not fighting back enough, but some military experts in our ranks recommended that we build stronger bunkers and more ammunition. The captains took their advice and so here we were. Now that the fifth town had been occupied, the captains were awakening to our need for action, especially since the fifth town was within easy range of us.

Our biggest weakness was that since our equipment was solar powered, it was inoperable during the night time hours. We had made many air tanks that could be changed once they were exhausted of pressure, but this took time and was not practical during an attack. Thus most of the enemy movements had been made under the cover of darkness. Our biggest advantage was that the United Nations' army was running out of gasoline and diesel fuel. Since there were no signs that oil was present on the planet, the United Nations was trying to convert to solar power. Most of their solar power equipment was stolen from the towns they conquered leading the civilians to adopt the practice of destroying their Dirt Demons and planes when they saw the enemy armies on the horizon. Of course the United Nations' growing dependency on solar power gave them the same night time weakness as the Republic had, and even more so since their soldiers seemed to have something against walking.

Another aspect of the warfare that is worth mentioning is that no side had adequate defenses against aircraft. The only thing the United Nations had was radar, which would only let them know that we were on our way. We had rapid-fire anti-aircraft guns in place, but their range was very limited, and because of their high rate of fire, they were practically useless at night since the operators would spend three times as much time changing air tanks as they would spend firing the guns. I suppose that they would save our lives in a pinch though. It was because of this aerial weakness that I suggested we launch an air strike against the United Nations' camps just before dusk. We could hit the camps closest to Gabville and while the occupational forces in Gabville were listening to their buddies cry for help on the radio, ground squads would invade Gabville and do a door to door search of the town and Gabville would be free.

The plan was accepted and we quickly set about getting ready to put in into action. We loaded PF2 fighters with bombs and loaded War Devils with ammunition and air tanks. We pumped up as many extra air tanks as we could carry. The United Nations would not expect us to make an attack so close to dark, so we would let the element of surprise work in our favor.

∞

Just before dusk, we took off. The sun was bright enough to give us almost full power. I looked at the two bombs strapped against the fuselage of my plane. Each one had the power to decimate an entire UN transport landing camp if I hit it dead center. They weighed five thousand pounds apiece and slowed the plane down considerably. I looked at my watch. In about fifteen minutes we would be over the target. There were two other planes in my formation. Over to the north another formation of three planes headed to another target camp. There were three camps near Gabville so we decided to send three planes to each one. With six bombs per camp, they would be eradicated no matter how much armor they had. The time passed quickly in the air. In no time I could see the camp up ahead. The transport was still in the center of the camp acting as a headquarters. They had built numerous buildings around it and had formed quite a large base. I kind of felt bad that we would have to destroy so much hard work, but they had asked for it.

I was flying at about three thousand feet. When I was almost above the camp I broke into a sudden dive. I aimed the plane for a group of buildings about two hundred feet from the transport and halfway between the transport and the edge of the base. Then I moved the plane a little bit towards the edge of the camp so that the vacuum of me pulling away from the bomb would put it right on course. I was one thousand feet above the ground when I armed and released the bomb. I pulled up and leveled out leaving the warhead to do its dirty work. The other two planes in my flight had slowed down before we got to the camp so that each would get there right after the smoke had cleared from the bomb before it. I was just about to climb for altitude when I heard the device detonate. I looked back to see half of the base enveloped in a rising mushroom cloud of orange flame. The shock wave moved away causing a

Phantasmagoria

ripple in the sand. I pulled the plane up sharply and shot for three thousand feet. A few seconds later I heard the report of the second bomb detonating. Finally I achieved altitude as the third bomb went off. I circled the camp once and then dove. The three shots had been aimed for the area around the transport. This time I aimed for the transport itself. I armed the bomb then released it. I leveled my plane out and banked over for Gabville. A few seconds later the bomb struck the transport, punctured it and detonated in the dead center of the craft, blowing it to smithereens. The fourth bomb blew a crater in the center of the camp where the transport had been and destroyed the underground bunker we had expected to be there. The last bomb vaporized the group of survivors trying to evacuate with trucks and captured Dirt Demons.

It only took me a couple of minutes to arrive at Gabville. I assessed the situation and quickly concluded that my assistance was not needed. The entire city was surrounded by War devils and troops openly marched the street providing support for teams that stormed buildings one at a time and cleared out the enemies. I flew low over the town, dipped my wings in a salute and headed back to our base. The sun was getting quite low and I had to supplement my motor with battery power. I knew that the batteries would last for a couple of hours at cruising speed, but I wanted to save as much power as I could in case we needed to evacuate the aircraft during the night. Airplanes were the most vulnerable when they were on the ground and so they would be the first to leave if enemy forces were seen in the vicinity.

It was about ten o'clock before the ground forces came in with the prisoners. They had suffered the first Phantasmagorian military casualty of the war. A young private was shot when making the first entry into a building. One guy kicked the door open and the others charged in while he got out of the way. The private that was killed was the first one in and took the machine gun fire. As soon as he heard the first gun go off, he started firing at the muzzle flashes and went down a hero. The next guy finished off the last hostiles and then they finished clearing out the building. We made a ceramic coffin and buried him under a twenty-one gun salute. We used captured M16's since our weapons were almost completely silent. I went to bed feeling sorry for the guy and his family. At the same time I also felt happy that he was our only casualty. Surely there would be many more; but we had only lost one guy while they had lost hundreds. It reminded me of the battle of New Orleans were Andrew Jackson beat the British. Only thirteen of his men were killed while the British lost over a thousand. The families of those thirteen then went and blamed Andrew Jackson for their deaths. I supposed that the family of the man we had lost today would be angry at us for not making him the last in line or something. After a while they would come to terms with the fact that he was a hero and it was an honor to die in the way he did, but until then we could expect to receive a little flak.

I lay in my bed for an hour thinking about the war and what we could do to try and cut it short. We had had great success against the little bases we had attacked so far. They were small and unprotected from an aerial attack. I wondered what would happen if we were to just go out one day and bomb them all to bits. We had the planes and bombs, so why not. I figured that somebody knew something that I did not or we would have already done it. The United Nations had not sent their troops here to be bombed to bits in a few days. They must have some sort of protection that kept the captains from making the decision to perform an all-out bombing mission and end them all in a few hours. It was almost as if we were playing some sort of cat and mouse game. We only performed local hit and run missions against each other. Even the attacks we had made tonight were only on targets in a small area and somewhat insignificant in the long run even though we thoroughly decimated the targets we hit.

I had heard rumors of a base on the other side of the planet. It seems that the United Nations had sent these crews to attempt to put down the opposition while they built a permanent base or bases on the other side of the planet where the only logical attacks would come from the air. Perhaps the captains feared that if we knocked out the local camps too fast then the main base would launch an all-out onslaught and destroy us before we could do anything about it. I figured that if the other base was that strong, then we needed to eradicate the local camps as fast as possible and assail the other base as soon as possible since they would do nothing but get stronger and more protected. I figured that an around the world flight would prove without a doubt whether or not a super base was under construction. I would volunteer for the flight in the morning. I fell asleep with very ambitious thoughts and dreams of battle.

∞

The next morning I ate my breakfast at the mess hall quickly and headed over to the office building where Ronnie and the other captains frequented. The base had been constructed out of premade buildings that we had put into storage in some of the numerous arsenals that were scattered over the area. There were eight of these bases in very strategic points between the towns. The base I was stationed at was situated on a high area surrounded by very rough hills. These hills would slow ground forces down as well as block low flying artillery shells. I power walked down the street and into the office building. About two dozen guys were sitting at desks working on paperwork. They had to keep track of every round of ammo and every piece of equipment on the base. They had to monitor the ratio of men building more planes and bombs to those on patrol and those on emergency standby while keeping production at the highest rate possible. No one would be able to take a break until the war was over.

I entered the conference room where Ronnie and a couple of the other captains were sitting at desks eating steak and eggs and talking about the day's activities. They were discussing launching a War Devil raid on one of the camps who had sent troops to investigate the bases we had bombed last night. They figured that since the base would have fewer men than usual, we could knock it off easily and then trap the other guys away from any fortifications. They reasoned that this setup would allow us to make a quick camp kill with minimal time out on the field. Ronnie filled me in on their plans then asked me for my opinion. I gave it to them.

"I think that we are only doing a fraction of what we could be doing. There are less than ten thousand enemy soldiers in the vicinity. While they outnumber us three to one, they are cowering behind those flimsy fortifications allowing us to bomb them to death with only a couple of airplanes per camp. Yet here we sit, fat and happy, not doing much about it. If we were to load up the planes, we could have every man of theirs killed by the time the sun goes down. We have the ability to end them all right now and all of you want to destroy one camp with War Devils. I say we fly for freedom right now!"

The captains looked at me in stunned silence. At last the captain of Phanta 4 spoke up. "We are mainly concerned that the United Nations is building a super base on the other hemisphere. We are fearful that by showing too much military strength would cause them to come at us with a force that no one could stop."

"So you are going to stand here and take them out one at a time while the super base builds strength and resources and the Phantasmagoria civilians think we are a bunch of chickens? If ya'll are concerned about a super base, then let me fly around the world and try to locate it. Once we know where it is, then we can knock off all of the local bases then fight it by itself."

Each of the captains present nodded. I do not know if they agreed that a trip around the world would be the most advantages thing right then, but at least they approved. Without another word I went outside to start getting supplies ready for my trip. I thought by the looks on their faces that they felt guilty about what I had said and just wanted to get rid of me for a couple of days. Whatever the case was, I was free to look for a super base if one existed. It would be logical that they would have one. I mean, why would they have ten thousand armed troops outside the civilized zone that were doing nothing? They had to be stalling for time. That was the only thing that made sense. Their troops could roll over the towns very quickly if they wanted to. Our planes would stop them, but they could take out most of the towns before we could get there. Maybe they did not want to destroy the cities. They could build a super base then retreat their numbers into that. They would tell the people to accept their government or face immediate destruction. If the people were frightened into submission, then they would be easier to handle than a people who were beaten in battle and made slaves. The frightened people would also be less likely to rise up against them in the future. That was the last thing I wanted to happen. We had to locate the super base and destroy it before they could finish armoring it and stocking it with weapons and supplies.

I picked out a battle tested PF2 and taxied over to the armory. I picked that airplane because it was new, but not so new that it could have a possible manufacturing defect. Well, it could always have defect, but the chances of that were slim. I loaded it with enough food and water to last me two weeks. I also loaded extra ammo for its Gatling guns. Satisfied that I had enough provisions to survive the trip, I taxied the airplane out to the runway and took off. I contemplated going and seeing Debra and Curtis at Seeone but decided that I needed every extra minute I could get. It had been two weeks since I saw them last and I really missed them, but time was valuable and I owed it to the citizens of the Republic to locate the super fort if it existed.

The day was warm and sunny as were all days on Phantasmagoria. I set my altitude to five thousand feet and put the plane at a comfortable cruising speed. The diameter of Phantasmagoria was the same as that of Earth, give or take a hundred miles, so I knew that a plane flying at two hundred miles an hour ground speed would go around it in about thirty-six hours. I planned to zigzag north and south somewhat to cover more area. I knew that there was no way one plane could cover the entire surface in any reasonable amount of time, but I knew of many areas that would offer strategic advantages from studying maps made by Earth satellites. I would make sure to cover these areas well and avoid the less likely places like open prairies. Chances are they would try to conceal it somehow or at least make it difficult to reach by land. To do that they would need canyons, swamps, and hills. They would also try to make it as close to the equator as possible so that they would be the same distance from our southern towns as our northern towns. Unlike Earth, the continents on Phantasmagoria were pretty much equally balanced between the northern and southern hemispheres. The location of our villages also reflected this. The Phanta crews had landed just north of the equator and so we had moved south more than we had moved north when we started building villages so that the population was equal.

I dipped my wings as a way of saying farewell when I flew passed the easternmost village. I felt like I was leaving civilization permanently as I soared away to the east. I knew I would be back in several days, but there was something about leaving that place behind that got to me. I shrugged off those feelings and turned to face the trip ahead of me. The terrain was pretty smooth here and could easily have been traversed by the jeeps and trucks that the United Nations used. I flew over a couple of their camps during the next few minutes of my flight. Either they were almost out of fuel for their vehicles and did not have enough solar powered vehicles, or they were stalling for a super base to be built. Those had to be the only reasons for them not overrunning all of our towns by now. I let the plane fly due east for a few more miles before turning it slightly north to examine some rough territory that might prove suitable for a base.

∞

I found the base on the second day of my journey. I was flying south southeast to check out some almost mountains that seemed like a likely place when I saw the super base. It was in an area that looked smooth and flat from space telescopes, but was actually surrounded by a network of thirty foot valleys and ridges. They probably picked the place thinking that it would slow down tanks and heavy artillery, but the problem was that we did not use any tanks. It would take time for a heavy armored vehicle to crawl up the ridges and they would have to brake hard going down them to keep from breaking something, but a War Devil would eat that terrain up. We could hide behind the ridges and fire up and over them then jump to the next one and repeat the process.

Phantasmagoria

I quickly increased my altitude to six thousand feet and began to circle the base. I pulled my spotting scope down from its clip above me and looked down to me right as I banked in my turn. To put it lightly, the base was huge. About two hundred buildings were there, surrounded by an armored ceramic wall that would take a lot of shelling before it came down. They had runways and artillery batteries. I could see several airplanes out on the tarmac and likely more were in the hangers. Out of everything they had, it was the artillery that concerned me. Row after row of guns ducked under ceramic armor. It would take quite a bit of bombing to punch through that armor. The number of guns meant that attack from the ground would probably result in many casualties. Perhaps if we used thermite bombs, the ceramic would crack under the intense heat and then be more easily destroyed by our high explosive bombs. Whatever we decided to do, we needed to do it before they made more airplanes and tried bombing missions of their own.

I turned the plane towards home and put it on top speed. The quicker I could get there, the more time we would have. It took me only a couple of hours to reach our base because I was flying in a straight line. I landed the plane and set off to find Ronnie and tell him the news. I found him at the office building talking with the Phanta 3 captain.

"Ronnie," I said. "We have to act fast. They do have a super base. It is about eight hundred miles south southeast from here. It is big! They have rows and rows of armor protected artillery and they are making airplanes. I think we can use thermite to crack the armor then bomb it, but we need to do away with the local bases first."

The two captains looked at me thoughtfully. Finally Ronnie spoke. "It will be dark in just a couple of hours. We will not be able to bomb the local camps today, but we can leave with the morning light. Let's get to work and get ready."

We left the office building and spread the word about the morning's attack. Couriers were dispatched to the other seven bases. Planes were loaded with bombs and pilots were assigned to go to different targets. We reviewed the flight plans and checked the planes over and over. Tomorrow we would launch the largest attack on the United Nations in their history.

∞

We rose before the morning light and were waiting by the planes when the sun popped up over the horizon. The sunrise lit up the sky brilliantly, creating a panorama of color. The colorful sunrise was like any other sunrise on Phantasmagoria, but just as the sun was rising, so was the Republic. We waited until about seven o'clock before taking off. By then there was plenty of sunlight to fly by and we could always use battery power if we were banking away from the sun. The entire strike force was divided into groups of three planes, each one carrying the usual two bombs. We had enough airplanes and trained pilots to be able to hit all of the known bases and still have a couple at our bases in reserve.

My flight group was assigned to one of the camps on the far eastern edge of civilization. We would be in the air the longest and our target camp would probably have the most warning time of any of the bases. Because of the warning time they would have, we needed to check and make sure that they had not evacuated any troops or trucks. We would destroy the base with three bombs, then search out any survivors and bomb them. After that, our instructions were to assist ground units in annihilating troops and trucks that were trying to escape. We passed several enemy bases on the way to our target. They were just sitting there, not knowing that their lives would be ended in a couple of minutes. I felt sorry for the men down there, but they were the ones that were threatening our freedom and way of life. I just hoped that they would not see us flying over and guess that a large scale attack was being made.

Finally our target base came into view. I was the second one to go in, so I slowed down and let the leader go ahead. I counted to twenty then accelerated back to my previous speed. I was approaching the dive point when the first bomb went off. About half of the base was consumed in the inferno. I dove and aimed for the

Phantasmagoria

side of the camp that still existed. I waited until my aim was just right, then I armed the bomb and released it. I quickly pulled up and leveled out. A few seconds later it detonated, showering my plane in debris. When I heard the second explosion, I looked back to see another fireball rising up in the air and casting a bright glow onto the surrounding terrain. I knew that no one could survive that pounding, but plane three had to add one for good measure. We searched the local area for any groups that may have been away from the camp. Finding none, we headed to the other nearby camps to help out. None of the local bases had any survivors, so we all joined together as a big group to fly back home. On our way back we saw a group of planes heading southeast, presumably to finish off some evacuees.

We were about five miles away from the base when we first saw the smoke. At first I thought that someone had tested some ammunition near the base, and then I realized that the base itself was smoking. I yelled and waved to the other guys and we shoved in the accelerator knobs. A minute later we were circling our own bombed base. I looked down in horror at the crumbled buildings and smoldering bodies. Not a single building had been left untouched. I motioned for us to land so we set our planes down on the sand nearby since the runway was destroyed. I barely let the plane stop before I jumped out and sprinted for the buildings. I could hear some ammo cooking off down the street, but I paid no attention to it and charged into the ruined base. Bodies lay roasted in the street and fires still burned on piles of rubble. The smell of napalm stung my nostrils as I surveyed the damage. Those planes we saw earlier must have been United Nations bombers returning to the super base. Dozens of U.N. aircraft lay around the base, victims to our anti-aircraft guns. The other pilots joined me. One of them was cursing vehemently and another was sobbing. My blood boiled. The napalm must have covered the ceramic and stuck to it, causing it to crumble under the intense heat. I knew many of the men who were now ashes. I sprinted down the street to the office building. That was when the first tears came to my eyes. The building must have taken a direct hit. I knew that there was no way that Ronnie or the rest of the Phanta captains could have survived that, but I had had to see anyway. I dug through the still hot ceramic chips towards where

The New Frontier

I thought the men might be. The ceramic chips burned and scratched my hands. Surely my friends must have been able to evacuate!

The first lump of charred flesh confirmed my fears, but I continued digging. The entire group of captains had been killed. I stood up, removed my flyers cap, bent my head down, and cried. Ronnie had been one of my best friends. He trusted me more than anyone did. He valued my opinion to allow me to join the decision making with the other captains. They all valued my opinion. And now they were all dead. I walked out of the rubble. This war was now personal. Before this I had fought the United Nations for our freedom and for a republic form of government. I had fought them so that our children could enjoy the right to bear arms, the freedom of speech, the freedom of travel, and the right to silence like Americans had enjoyed for almost three hundred years. But now the United Nations had taken the lives of my friends. They had destroyed the lives of some of the best men I had ever known. I would now fight to avenge their deaths. Before I climbed out of the pile of ceramic rubble, I turned around and looked at the charred flesh lying there on the ground. "I will not allow your deaths to be in vain," I said. "I promise that the ones who did this to you will pay for their deeds. I will personally make sure that your mission to drive the United Nations from Phantasmagoria will be completed." I paused for a moment. "They will pay for this," I mumbled quietly to the dead men. "Oh boy will they pay!"

∞

It took several hours for the bodies of the dead to be buried. We gave each man his own grave and when we were done, we fired a salute over their graves. Nobody said it outright, but I could tell that they all looked to me as their leader now that the captains were gone. We had never officially said that the captains were in charge of the military, but they started the rebellion and took charge of it so we looked to them as our leaders. Now they were gone and I was the only one that they considered to be an equal. Now the men looked to me for leadership. I knew that this was my opportunity to avenge the deaths of my friends. I could not destroy the United

Phantasmagoria

Nations by myself, but I could command the force that did and that would be the same thing.

While the men worked on locating the families of the deceased, I hopped on a PF2 and headed to Seeone to give the news to the captains' wives. I decided that it would be best if I told them in person instead of letting them find out when everyone else found out. I found the ladies in a small restaurant eating lunch. I approached the table with my hat in my hands and waited to be acknowledged. Debra was the first one to notice me. She jumped up with joy and ran to throw her arms around me. I gave her a quick hug, and then pushed her away. She looked surprised.

"Flynn, what is the matter?" she asked. "You look terrible!" The other ladies stopped talking and looked at me inquisitively.

"Ladies," I began in a slow, sorrowful voice. "I am sorry to have to be the one to bring you this news, but there was a disaster at the base today." They gasped and looked horrified. "They are all gone!" I burst out, unable to control the boiling passion and sorrow inside of me. "All six of them are gone! We were out bombing all of the local camps. While we were gone airplanes came from the super base and destroyed all eight of our bases. Every one of the Phanta captains was there and now they are all dead!" Each of the ladies sat there, unbelieving, before bursting into tears. "I swear there was nothing we could do about it. We had no idea that they had any airplanes with those capabilities. Even if we did they probably still could have bombed us because we do not have any radar system. There was nothing we could do!" I sat down at the table next to the ladies and laid my head on me arms, sobbing, trying to get the images of charred comrades out of my head.

Ronnie's wife looked at me through tear streaked eyes. "Kill them, Flynn! Kill every one of those tyrants!"

"I will," I promised. "I will."

∞

I woke up with the morning sun the next morning. We had to sleep on the ground because our base was destroyed and the barracks of course went with it. I got up and walked over to the stack of food containers and got out some dried meat. I was busy gnawing on it when a couple of officers approached me. I stopped eating and asked them if I could help them.

"Commander," began their spokesman. "Are we fighting for a constitutional democratic republic in this war?"

"Yes sir," I replied. "The constitution of the Republic of Phantasmagoria states that the government will be run using democratic methods."

"The constitution also gives provisions for the people to elect their leaders by a majority vote."

"That is true," I said, not sure what he was up to.

"Well, before yesterday's bombing we always looked to the captains of the Phanta missions as our leaders."

"I also looked up to them as my superiors."

"Yesterday afternoon we held an election. We sent couriers to the different base sites and towns and gathered a vote."

"I see," I said, somewhat curious as to what the people had decided.

"The people voted you in as commander-in-chief. Congratulations commander!" The spokesman shook my hand while the others saluted. I was too stunned to speak. I shook the man's hand and saluted the others. I had expected to provide leadership for the military, but I did not expect to be made official commander-in-chief. The group's spokesman pinned five gold stars onto my ragged clothing. I felt like General Washington must have felt when he looked at the continental army. My men wore ragged clothes that were often partially shed in the sweltering heat. Much of our supplies had been destroyed in the bombings. And they were all looking to me for leadership and I would lead them to victory like Washington had done three hundred years ago. There was not much difference between now and then. Both Washington's and my armies were made up of desperate men who wanted to live freely in the way they saw fit. We were also both fighting tyranny and the strongest military forces of our times. The only difference between us was that my army could manufacture its own resources.

I paced around over the dunes near the rubble and thought about our next move. We would have to get to work on making more ammunition as fast as we could. The super base would know that we were staying near our bombed out bases and could come and bomb us too at any time if they wanted. We needed to get all of our equipment that was still intact out of there. The only thing I could not figure out is where we should go. We had found out the hard way that the United Nations could bomb our permanent bases if we had any. We would need to separate and operate in a nomadic style. We needed to build transport buggies to carry portable ceramic shelters. We could clip the walls together at night and then move them during the day to keep the United Nations guessing.

I told the men what we needed to do, and they started working on it. The first thing was to see how much of our equipment could be salvaged. Most of the War Devils were destroyed, but we still had most of our bombs and small arms ammunition because a lot of it was stored underground and in satellite locations. I instructed some of the men to make the transport buggies that we would need. We organized the bombs and ammo and counted our losses. We still had most of the equipment that we needed to make the ceramics and smelt metals. Because of that we would not have to start completely over from scratch. By three in the afternoon we were packed up and ready to go. Planes had been going to and fro from the other bases bringing reports. The effects on the other bases were varied. Some had lost their ammo while others had lost only their machines and yet others had lost a combination. Whatever the case was, I gave the same order: patch up what you can and then split into platoons of a hundred guys or so and just try to disappear into the wilderness. Let the other groups

The New Frontier

know where you are and we will let you know what to do when we plan attack missions. Build only as much equipment as you can conceal easily.

We were going to go under cover and stay concealed only until we could replace the stuff we had lost. Then we would coordinate a huge attack against the United Nations' super base.

∞

A week later we were still in hiding. My platoon was made up of some of the best soldiers that we had. Even though I had been elected commander-in-chief, I still worked and lived with my squad. Our platoon, like all of the other platoons, had moved to the north and south of the civilized zone. That area was referred to as wilderness and in fact it was. Except for us, it was devoid of humans. My platoon had hid out in some small canyons where we found good iron mining. We made loads of thermite and loaded it into bombs and grenades. The United Nations had used napalm against us and it crumbled the ceramic. Thermite would burn much hotter and would probably reduce their armored bunkers to sand like it had been originally. Right now the main task was to simply lay low and stockpile ammunition. We had left our arsenals in the civilized zone alone so that they would be there if we really needed them. We then made new arsenals and stockpiles outside the civilized zone.

I had been busy all day making thermite, so that night I took a walk to clear my head. The sunset was beautiful as usual. I found a nice boulder in a high place and sat on it to watch the sun continue its descent. I thought about what the United Nations might be doing right then. They probably thought that we had either disappeared to regroup or that we had disappeared to skulk in the woods about our first defeat. In reality the first was true. We were out to regroup, but in a way like no one would ever expect. We had started the war to gain freedom and independence from Earth, and now we also fought to avenge the deaths of those who had given their lives for the cause. The men who had been killed in the base bombing had died trying to gain freedom from tyranny so by golly we would complete the task. I knew that attacking the super base would be a difficult task and that even if we succeeded in destroying it, there was no guarantee that the United Nations would give up. After all, they could not return to Earth until their mission was complete and transports were sent for them. But then there was no guarantee that they would keep fighting. All I knew was that as long as the super base was intact, we would still have to deal with the United Nations. I thought about what I had seen of the base. I had only seen about a thousand men in there. No doubt they were relying on the local forces that we had decimated. The fort could easily contain ten thousand men and their weapons. I was sure that they had already added more armor and artillery to the massive defenses that they already had. Why would they not? The resources were free for the taking and they had nothing but time. Besides, they knew that we had some pretty powerful bombs which could punch through quite a bit of armor.

I pondered where they got napalm from. We had no known petroleum sources and as far as I knew the United Nations were running low on gasoline. It was likely that they had used biofuels of some sort. Whatever it was, we would have to watch out for it if we made any sort of ground attack. Napalm was a very sticky substance that was almost impossible to put out once it was lit. Some ground forces on Earth had tried covering themselves with foam when they expected napalm to be dropped. The foam gave off carbon dioxide which suffocated the fire. The downside to this was that the equipment used to produce the foam was bulky and heavy. The soldiers also had to wear oxygen masks or the foam would suffocate them too. This made it hard for the soldier to move around and lowered the amount of weaponry that a soldier could carry.

The best way to combat napalm was to never let it hit you. The best way for us to achieve this was for us to shoot down the airplanes dropping it. Tracking missiles were pretty much out of the question because of their complexity. The flak cannon would probably work, but at a short range. I had had the guys build a simple flak cannon and test it against a dummy drone. The flak worked just fine. We guessed that the pilot would be most vulnerable to the flak because the planes that both sides used were open cockpit. I also thought that perhaps an electromagnetic pulse would be an effective weapon against their planes and other solar powered vehicles. A powerful pulse could fry a solar panel and electric motor. The problem was that the equipment could potentially be bulky and we ran a high risk of frying our guys too.

Another way to avoid napalm was to destroy their planes on the ground. An airplane is the most vulnerable on the ground and quite sensitive. It would not take a very big bomb to knock all of their planes out of commission. We could then move in and attack them at our leisure. All of these plans would need to be thought about and stewed over for many hours. A warm breeze blew over me, flapping my ragged clothing. Perhaps we could use the weather to our advantage. I could not imagine any base of theirs having more than a day's supply of water on hand at any one time. In this heat a thousand men drank a lot of water. If something were to happen to their supply there would be a panic. The human body could only live for three days without water and that statistic was assuming that the person was performing limited activity and in perfect health. Add sweltering heat and a combat environment and that number would fall drastically. I knew of no way to remove their water supply at the moment, but it would definitely be something to keep in mind.

I looked down at our little temporary base and the men milling around it. Those men looked up to me. They trusted me with their lives. If I told them to go somewhere, they would go there happily and without question. If I told them to shoot at something and take a base or location, they would shoot at that thing and capture it no matter how fortified the thing was and no matter how impossible the task seemed. It was not necessarily that they trusted my abilities, but they trusted my wisdom. If something seemed illogical, they would accept it as fact while thinking that I knew something that they did not. It was not that they were sheep, but it was that they had phenomenal amounts of faith. I was not the most knowledgeable person when it came to military tactics, but I had seen plans work and I had seen plans fail. I took this knowledge and logically combined it with advice from others who had more combat experience than I did. I would then create a plan. One who has a plan of action is infinitely more ahead of one who simply reacts to his surroundings even if the plan is slightly flawed. I knew this and combined this power with the power of goals and the power of determination to lead my men. By using such stratagem, I had earned their respect and trust.

On the other hand, I had learned that the United Nations very rarely had such advantages. While they would not operate without a plan and goals, their leaders did not have the trust in their men like I had. Their soldiers came from all over Earth and were forever transferring to different platoons. This made it very difficult for a commanding officer to earn the true trust of his men. On an ordinary battlefield this might not have very obvious consequences, but this was no ordinary battlefield.

I climbed down from my rock and headed into camp. Our temporary buildings had been covered in rocks on their roofs to help make them harder to spot from the air. I strolled over to the bunker that I shared with some of the men. I had made it a practice to sleep in the same accommodations as the privates did and eat the same food. I found Lawrence waiting for me by the door.

"Hey Lawrence," I said. It had been quite a while since I had been able to talk to my buddy.

"Flynn, I have been thinking." I knew that when he started talking this way he was probably about to propose something drastic, but I let him continue. "I would like to go to the super base with some of my other sniper friends and try to pick off some of the leaders."

I mulled this over in my head for a minute. To do something like this would be very risky, but if they were successful it would be a great victory for the Republic. My sensitive side told me not to let my buddy do anything dangerous, but my realistic side told me that it was a soldier's job to risk his life for his cause and that this would be a great chance for Lawrence and his sniper friends. My commander side told me that the lives of a few rogue generals were no trade for the lives of even one of my men, but my optimistic side told me that even if Lawrence and his friends were killed, they would probably be able to take out a couple dozen officers before being shelled with artillery from the base or napalmed from the air. "Lawrence, as a friend and your commanding officer I will inform you that your lives are not worth the lives of whoever you might kill in the base, but as a military leader and a soldier I will tell you that such a mission will be of great value to the Republic. I know that you and your friends are willing to give your lives for the Republic. Because of that I am going to let you go if you want to, but remember that no one will hold anything against you all if you decide not to go. The decision is yours."

"Thank you Flynn," he said shaking my hand and saluting me. "We leave tomorrow with the sun."

"Lawrence," I said. "Take care of yourself. I do not want to have to tell your wife anything."

He nodded soberly and walked off to his barracks. There goes a good man, I thought. A really good man.

Chapter 11

Lawrence left the next morning bright and early just like he said he would. He took six other snipers with him and a plane pilot. Something in my gut told me that he would be just fine. They took plenty of ammo with them and a detailed map of the base.

My description of the base had been updated several times by other flybys. The pilots would strap on oxygen tanks and fly at ten thousand feet or more to escape detection. Usually they would go with a pilot and a spotter and circle the base for several hours while drawing detailed maps. These maps were brought back and compiled together then distributed to all of our bases. Because of this, Lawrence and the other snipers would have a good idea as to where the base's commanders would be hanging out.

With most of the leaders gone, it would be easier to launch more formal attacks on the base. There were certain things that would probably happen within their ranks that would distract from their operations. First there would be lots of promotions. If someone did not like his commanding officer and the way he ran things, then he might try to make changes and changes would break up the flow of things for them. Second, having an equally ranked officer suddenly promoted and put in your command could create a number of private feuds. Nobody wanted their buddy and equal put in charge of them. None of these things were guaranteed to happen, but by knowing human nature I could assume that they would. It was things like this that reduced the efficiency of a fighting force. Because of this, when someone was given a promotion above his peers, he was usually transferred to another area. I was betting that things would happen so fast there at the super base that they would just promote the next guy in line. Even if they did not, the simple fact that there were promotions because someone was eliminated would cause jealousy. A little jealousy could go a long way.

I watched the sniper crew disappear into the sky. They would be returning tomorrow if everything went well. If it did not go well, they would never return. I tried to shrug the worry from my mind that day as I mixed thermite. I had never been a worrier, but I really did not want to tell another wife that her husband was not coming home. I decided that the best way to not worry about the snipers was to immerse myself in my work.

Right now I was mixing the thermite that we would drop on the super base. Thermite is made out of iron oxide and aluminum oxide. When the iron oxide ignites, it burns at several thousand degrees and will eat right through metal armor. In the case of the super base, the armor was very hard ceramic. Of course, the harder ceramic got, the more brittle it tended to get and that is why we had decided to hit it with an incendiary weapon first. We were making thermite bombs to be dropped from planes as well as artillery shells which could be shot from our various assortments of howitzers and grenade launchers.

After stirring thermite together for several hours, I gave the job to someone else and went outside to check on the rest of base operations. We had built a strong fleet of War Devils and transport trucks. We had also built quite a large force of PF2 fighter planes. I could see planes circling up in the sky; mainly being used for training. I had wanted as many people to be trained as pilots as possible. The more pilots we had, the better off we were. I checked out the building that we were using as an ammunition factory. The guys in there were working their tails off making little round bullets. Satisfied that all was in order, I headed off to the mess hall for a well-deserved meal.

∞

About two in the afternoon the next day I started receiving reports of a plane on the horizon, coming from the direction of the super base. I hoped that it would be the snipers because they were due in, but I still ordered a squadron of planes dispatched to escort it in and I also ordered the anti-aircraft guns armed and loaded. We had set up radios in some of the planes so that they could communicate with the base when they were on patrol. They only transmitted on a very low power, of course. I walked out to the runway so that if it was the snipers returning, I could be the first to speak to them. Several minutes later the escort appeared. They were moving very slowly, surrounding a very damaged PF2. I recognized the tail number of the one that the sniper team had taken. The good news was that they had returned, but I hoped that all of them were alive and okay. The PF2 made a slow, careful landing on the runway and limped over to where I stood. I waited for them to stop, and then I hustled to the fuselage. Lawrence jumped out and shook my hand.

"Are you all okay?" I asked, thrilled to see them back again.

"Yeah, we are okay," Lawrence said in an excited voice. "Harry was wounded in the leg, but he will be just fine." A couple of other guys helped Harry out of the plane and took him to the barracks for treatment of his wound. Lawrence and I walked towards the mess hall for a snack and a drink while he filled me in on the details of their mission.

"How did it go?" I asked.

"The mission was a grand success. We spent yesterday afternoon identifying the leaders by their uniforms and trying to figure out where they hang out. It was not very hard actually. Then this morning they happened to have some sort of party in the officer's mess hall. We had Chad drop us off at different points around the base. You cannot see it from the air because the perception is off, but they are really in a slight valley. It looks like the area is level, but I think it is an optical illusion from the grid work of ridges. Anyway, they seem to think that they are pretty well protected by all of those guns that they have aimed at the hills. Because of that, they went and built the officer's mess hall at a high point in the base with large windows so that they could look down at the base and see it in operation while they ate. However, those large windows are what snipers love. Each one of us was about two thousand yards distant from the mess hall. We waited until about eight in the morning when the place was just chock full of people. Then at precisely eight o' five, we opened up on the place. We just put those babies on full auto and let them have it! You could just watch row after row of them collapse on the tables and the floor! It was both terrific and horrible at the same time." I laughed and nodded understandingly. Then he continued his narrative. "It took quite a bit of time before they realized that the officer's mess hall had been fired on. After all, there were no gunshot sounds so the only sound they heard was the yelling and breaking of glass." He paused as we entered the mess hall.

"I swear I could hear them screaming as they went down even at the range I was at," he added soberly. I just nodded and put my hand on his shoulder. It was very hard to hear the screams of men dying, especially when you knew that you were the one who had pulled the trigger. "Anyway," he said shaking his head as if to try to shake the memory out. "Once the mess hall was completely riddled with bullets, I replaced the mag in my gun, switched it to semi-auto mode and started picking off the guys running for the artillery. I did not worry about the ones actually using the guns as much as I did the ones yelling instructions. Once they got to the guns, it was kind of hard to hit them because of the way that thing is built, but I still managed to hit a few." The base had been built on a

Phantasmagoria

hill so that each bunker could be right behind the next one and higher than the one in front of it as well. This made the bunkers look like a solid wall of guns and for all intents and purposes, it was.

"Then what happened?" I asked as we sat down with some doughnuts and coffee. The hydroponic greenhouse-grown coffee tasted just like its Earth-grown counterpart. "If all of this happened at eight in the morning, why are you guys just getting back?"

"Well, they opened up on the hills with those howitzers and machine guns. We had agreed that when that happened we would all run down the back of the ridge we were behind and Chad would come and pick us up with the plane in our turn. Harry got hit in the leg as he was going, but still managed to get down behind the cover. Chad would pick up a guy, then fly to the next one and pick him up. The plane got hit several times as you could see, but we managed to avoid getting hit. After he picked all of us up, we got some altitude and started circling the base." He paused and took a bite of doughnut. "I guess I will tell you now," he said. "After we left, we stopped at a camp to the east of us and picked up two thermite bombs. It was not part of the original plan, but we figured that if we were going that way we might as well test them out and see how well they worked against their ceramic armor." He paused and took another bite, probably trying to figure out how I would take the information.

"Go on," I said. "How did they work?"

"The fort crumbled right on top of their heads and poured the thermite right on them, burning through the next layer and the next. Each bomb crumbled three layers of fortification."

"You know I could have all of you court marshaled for testing without permission, but I am thrilled that you all were able to do that," I said, grinning. "Now we know our limits and abilities. Of course they do too, but there is not much they can do about it." I finished my coffee and stood up. "This mission has been the turning point in the war. Now they are leaderless and they also know that our bombs can cook right through their armor. We will let that sink in for a couple of days and then will bomb them right off the planet."

"Why not get ready and bomb them tomorrow?" Lawrence asked, standing up and following me outside. I glanced around at the busy camp before answering.

"When they realize how vulnerable they are, they will be frightened. They will have no leaders to lie to them and tell them that everything is okay. We will let fear work for us for a few days, and then we will let them taste our rebel power."

∞

I laid flat on the Phantasmagorian soil looking at the super base through my spotting scope. I watched the United Nations soldiers going on with their normal daily duties. Some were scrubbing the large guns that provided ground protection for the base. Mechanics worked on airplanes sitting out on the tarmac. I saw some officers having a fist fight, presumably about who was in control. Since Lawrence and his friends had shot most of the higher ranking officers, the men had spent most of their time fighting for control. Evidently they had not contacted Earth for instructions regarding promotions. I saw a large team of men hard at work on the side of the base where the sniper team had dropped the thermite bombs. The soldiers had almost finished the fortifications, but the places were only protected by temporary guns which had been rolled into place. I glanced down at my watch. The bombers should be here any minute. They would lay waste to the base then ground troops would move in and finish them off. The base was almost completely surrounded by rebel forces ready to attack when the smoke cleared. To my left and right were men laying on their stomachs watching the base and waiting for the bombs to strike. As soon as I gave the orders, the men would run back to the War Devils parked thirty feet behind them and we would advance for the final blow. Once the base was completely laid to waste, I would have a radio set up to communicate with Earth and discuss terms of surrender.

Phantasmagoria

My men had been planning and preparing for this day for nearly a week. We had scanned the entire planet multiple times for other super bases, but this was the only one. Destroy this base and the United Nations would have no more troops on the planet. They could always send more, but we would do the same thing to them as we had done to the others. I looked into the eyes of my men as they laid on the Phantasmagoria soil. They could taste the victory in their mouths. The last thing between them and the freedom to live their lives in the way they chose was right in front of them. Once this one base was reduced to rubble, they could return to their families and life would be normal again.

Since almost every male colonist was here fighting, the women had to do everything back in the colonies. This had both positive and negative effects. The women had done a great job of keeping everything in order, but it was hard for the women to take care of towns that were built for twice the population that they had. Besides, most of the women were not engineers or scientists so they could not build new planes and buggies that they would need. Once this war was over, the men could return and the balance of life would be normal again.

At exactly nine o'clock I heard the whir of two hundred airplanes. Each one was almost completely silent, but collectively you could hear them approaching. I looked through my spotter scope. The men at the base heard the roar but did not know what it was. Some of them crouched behind the guns for security like a child and its blanket, yet others stood there and scanned the sky and horizon. I looked over my shoulder to see the huge formation coming at about three thousand feet. They were spread out in such a way as to make it look like there were a thousand aircraft. I looked at the fort again. The men had seen the planes and were panicking. Some fired harmlessly at the planes with their howitzers. Others ran around in circles while a few took the coward's way out by shooting themselves. The base was a perfect example of complete chaos.

The formation of planes flew over the fort. A couple of predetermined planes dove down and let their thermite bombs soar towards the base. They dove and released their bombs at the same time so that the bombs would strike the base and detonate at the same time. This would reduce the chances of one bomb going off and shooting a nearby bomb out into the hills to explode. As soon as the planes let their payloads go, they leveled out, and then climbed to rejoin the formation. The bombs struck the ceramic armor and burst, covering the ceramic surface of the base with fiery thermite. The exposed pieces of ground between the interior buildings were fused into glass. The buildings crumbled like fine china or eggshell. The steel frames of their aircraft melted and boiled in puddles on the ground. The artillery bunkers crumbled into sand and collapsed onto the guns beneath them. Grenades and artillery ammunition cooked off beneath the inferno and sent their projectiles hurtling wildly through the powdered ceramic on top of them. Cooked-off bullets popped out of the flames making the collapsing bunkers seem to boil. The walls of the base fell like those of Jericho as the firestorm continued. Men shriveled in the heat like raisins and turned to ash before their bodies could hit the ground which fell out from underneath them.

I stared at the flaming firestorm in shock. I was glad to be rid of the base, but seeing it go like this was intense. I could see the formation returning. This time they were going to drop high explosive bombs. These were supposed to scatter the thermite and the powdered ceramic armor allowing them to thermite the layer below. They would continue this until they ran out of bombs or the base was nothing but a smoking crater. Five planes dropped from the formation and dove towards the base. The pilots could probably feel the rising heat in their faces. They let go of their munitions and rejoined the formation. The bombs detonated and sent the thermite flying. We were about a mile away so we did not get hit by any of it, but some came pretty close. The plan had worked perfectly. New layers of the bunkers were exposed to the sky. The center of the base was thoroughly destroyed so the bombers centered their bombs on the edges. The planes turned around as a group and came back for another run. The thermite bombs crumbled the newly exposed bunkers just like it had done the first time. The planes came back for three more passes before heading home.

Phantasmagoria

With a wave of my arm the men ran to the War Devils and mounted for the final charge. There was probably nothing else for us to destroy, but we wanted to make sure. This would also let the men who were not chosen to fly bombers to feel like they had participated in the action. I climbed into the shotgun seat of my squad's War Devil and we took off for the base. The ground was littered with flaming piles of thermite-covered ceramic. As it cooled, most of the ceramic fused together forming a porous lump. I could see the War Devils moving in from all angles. We were all riding in charge position with weapons ready and the flame thrower operator riding on top in the field howitzer seat.

We were about two hundred yards from the out skirts of the base when the unexpected happened. The entire base seemed to rise up with a series of small explosions. Debris flew past and landed on all of the War Devils. Lawrence slammed on the brakes and we slid to a halt with the rest of the force. In front of us where the base had been was a flat plate of steel and ceramic armor. The explosives had cleaned it of the rubble from the super base above it. All at once large square manholes opened up and men jumped out of them with smoking machine guns. During the split-second that everything happened I had brought my grenade launcher up into firing position. I lined up my sights on one of the manholes and fired. The grenade shot out and flew past the men into the underground shelter. Lawrence grabbed up his assault rifle and rolled out of his seat, firing. Frank hit one of the open man holes with the field howitzer, sending bodies flying. I rotated the cylinder of my grenade launcher and fired another grenade. Explosive-filled machine gun bullets riddled the front of our War Devil. Jacob jumped out and lay down beside the buggy in a prone position with his assault rifle. On either side of us were the other War Devil teams, each one pumping its full firepower into the unexpected bunker.

The firefight continued for about five minutes. Finally, the shots stopped coming from the underground shelter. We continued to launch explosives at it for a moment, and then we stopped as well. I could hear the moans of the wounded from both sides. I reloaded my grenade launcher then set it down and grabbed my assault rifle. I climbed out of the totaled War Devil and surveyed the scene. War Devils lay wrecked all around the shelter, some with bodies lying in them. Beside the War devils lay men with automatic rifles trained on the open manholes. Bodies lay near the manholes with arms outstretched. Guns lay in puddles of blood, their magazines empty. I waved my hand towards the shelter and the survivors of my men rose and approached the shelter.

A couple of the War Devils had managed to escape the assault so we rolled them to the man holes and tilted them forward so that we could use their headlights to see in. The floor was strewn with bodies and weapons. A group of soldiers stood in the center with raised hands. We ordered them to exit and they did. We closed the manhole covers and fused them shut with thermite after giving anyone hiding in the dark corners a chance to surrender.

It was then that I felt the pain in my leg. I looked down to see a whole in my pants just below my knee and blood draining off the side of my shoe. I pulled my bowie knife out of its sheath and sliced my pants open all the way up to the knee. Sure enough, I had taken a bullet through the leg. Some of the guys nearby noticed and came to my aid. One of them checked to see if the bullet had hit the bone while another tore his shirt into strips for a bandage. The bullet had missed the bone so I knew that I would be okay unless infection set in. They wrapped my leg up tightly and then once the bleeding stopped they put some ointment on it to fight infection and wrapped it up in clean bandages. One of the guys had broken part of the frame off of a totaled War Devil and gave it to me for a crutch.

We had lost about twenty-five men in the cross-fire. Since the men inside the shelter could not see us, they had come out of the manholes spraying and praying instead of aiming. That and the explosions that they had used to clear off the shelter had given us the chance to ready our weapons, as short as it was. I looked at the prisoners. They were all men who had either been wounded and fell back or men had ran out of ammo and moved back to get more. I walked over to where some of my men were setting up a radio for me to talk to Earth. They set up the solar panels and the antenna.

Finally, it was done. I sat in the chair in front of the set and turned on the set. I was going to use the frequency that we had used to talk to Spacex back when we were still with the Phanta missions. I selected the frequency, listened for a clear signal, and made my call.

"This is The Republic of Phantasmagoria calling Earth. This is The Republic of Phantasmagoria calling Earth. Come in, Earth." I set the mic down. The soldiers near me had stopped what they were doing and listened. Not hearing anything, I repeated the call. After the second time calling, I got a response.

"Republic of Phantasmagoria, this is Spacex mission control. How can we help you?"

"Spacex, how quickly can you get representatives of the United Nations to your location? We have destroyed their base on our planet and we demand that the organization surrender. We would like to discuss the terms of that surrender."

There was a moment's pause. "Hold on Phantasmagoria. We have notified the United Nations headquarters in New York City of your demands and we are trying to hook up an internet connection between them and our radio so that you all can negotiate directly. This may take around fifteen minutes."

"Thank you Spacex." The rest of my men had gathered around to listen to the most important conversation in the history of Phantasmagoria. About ten minutes later the United Nations came on.

"Phantasmagoria, this is the Secretary-General of the United Nations. Who is this speaking for the Republic?"

"This is Flynn Carson, former pilot of the Phanta 1 Spacex mission and now the commander-in-chief of the Republic of Phantasmagoria's armed forces."

"Well, Commander, I have received a report that you want to negotiate the terms of our surrender. What makes you think that we will surrender to you?"

"I think that you will surrender because it is the wisest thing to do. We have destroyed all of your forces on this planet. Right now I am sitting on top of the foundation of the super base that your forces have built here. If you send more troops, they will receive the same treatment. If you send nuclear missiles from Earth, we will intercept them and destroy them. All we want is to be able to govern ourselves with a republic. We ask that the United Nations step aside and let us do that."

The secretary-general was silent for a minute. "Our space telescopes have confirmed the destruction of our forces, but as the secretary-general, I cannot make the surrender myself. It will require a majority vote of the General Assembly."

"How quickly can you get the General Assembly together?"

"It would take several days to get them together in person, but we can vote on it virtually. What terms do you offer?"

"Our terms are as follows: The United Nations will have nothing to do with Phantasmagoria. If you send ambassadors or troops here, they will be tried for high treason, punishable by death. The United Nations will also not interfere with any Earth-based government and their dealings with us. You will step out of the way and allow the United States to continue to send colonists and will allow any other nation to send colonists if they want. The United Nations will recognize the Republic of Phantasmagoria as a separate nation and independent of all international laws and treaties. That means ALL treaties, including the Geneva Convention Treaty. The United Nations will also pay for war damages to the Republic of Phantasmagoria to the amount of six billion US dollars in the form of copper, nickel, zinc, titanium, aluminum, and uranium. These payments will be made on a ten year agreement with an interest rate of six percent compounded annually. Those are our terms."

"I will present those terms to the General Assembly and we should have an answer in less than two hours. Can we contact you on this frequency when we have our answer?"

"Yes, we will be listening." We gave the usual formal goodbyes and I put the microphone down. I had been thinking about the terms for a couple of weeks and they sounded pretty fair. I just hoped that I had covered everything that needed to be covered. If loopholes arose, then we would take care of it through negotiations and if need be, military power. I had no qualms about attacking United Nations' forces on Earth if that is what needed to be done. The fact was, the United Nations had overstepped their bounds and now they were going to make up for it.

I had chosen the war damages to be paid in metals because we needed metals. Sure we had had success smelting iron, copper, and gold, but we needed more copper than we could smelt and we also needed the other metals that I had specified. The uranium would be optional because we always had solar power, but there were some things that would be nice to be able to operate at night.

We radioed the pilots who had returned to our camps and told them what had happened. They would outfit PF3's with extra seat nets and come pick us up. I looked at the shot-up War Devils. A couple of them were salvageable, but we would probably strip most of them of their motors and solar panels and then use them as targets. We still had a long way to go in making bullet-resistant ceramics. The machine gun bullets had done quite a number on the War Devils. We had an area of improvement that some of the engineers and chemists could sink their teeth into when we got back and everything was normal again.

Once we were back at the bases, I told all of the guys to pack up and get ready to go home. I wanted to let them celebrate, but I figured that they would be better off celebrating once we returned to our families. They set to work with energy and soon had all of the buildings disassembled and loaded onto the trucks. I sat down in a chair in the shade to rest my leg. From time to time the guys would stop by to see if I needed anything. I looked at the other wounded men laying down in the shade around me. Some only had flesh wounds like me, but others were in critical condition. I had wanted them taken to the villages for better treatment directly, but somehow the order had gotten mistaken and so they were here, but the PF3's were being readied for them.

One man in particular touched my heart. He had been shot in the right shoulder and had his arm blown off. He was lying there with his eyes closed, hovering between life and death, but somehow he managed to find the strength to salute me with his left arm when I got up to get a drink. I knelt down beside him, put my hand on his forehead and said a quick little prayer for his recovery. I patted him on his good shoulder and went to get my drink, blinded by tears.

About three o'clock we got a reply from the United Nations. I was just about to climb into the backseat of a PF2 for the ride to Seeone when a guy ran up and told me that the Secretary-general was on the radio and wanted to speak with me. I grabbed up my crutch and hobbled to the radio set. I picked up the mic.

"This is the commander-in-chief," I said.

"Hi Commander," said the Secretary-General. "To get right to the point, the General Assembly decided that your terms of surrender were satisfactory and we agree. On the next transport from Earth will be a signed copy of our agreement for the purpose of records."

"Thank you very much Secretary-general," I said. "Now if you will excuse me, one of your soldiers shot me in the leg and I also need to go see my wife whom I have not seen in over a month."

I hobbled back to the plane and climbed into it. The pilot was about to take off when Lawrence ran over waving for him to stop. Lawrence climbed in then told the pilot to continue. I smiled at my friend. "I guess it is all over," I said as the aircraft rose into the air.

"Well at least the war part is," he replied, looking down at the ground which seemed to fall away as the plane rose higher and higher. "We still have a society to build."

"Yes," I said. "That is true. However, we already have a good start on that." We sat in silence for the rest of the trip. We both missed our wives and were anxious to see how they had been getting along. Well, at least it would only be about thirty minutes before we would be able to see them. I leaned back in the hard, ceramic seat and closed my eyes to take my first nap during the day in six months.

Chapter 12

I was awakened by the jolting of the plane hitting the runway. I shook my head and looked over at Lawrence. He was just waking up too. Thankfully our pilot had stayed awake. He taxied the plane to the parking area of the tarmac. I climbed out, grabbed my cane, thanked the pilot for a smooth flight, and started limping down the street to find Debra. Lawrence joined me and we walked together. The PF3's that they had used to transport the wounded soldiers were there, but those were the only military planes. The streets of Seeone were somewhat crowded with women going about their business. It was three o'clock in the afternoon.

Lawrence and I walked into the restaurant which had been the hub of information in the town. A waitress walked over and asked us politely if we wanted a booth or table. I told her that we were just back from war and we wanted to know where our wives were. She eyed our filthy ragged clothing, but tried not to show her disgust. She said that the girl behind the counter wound know more than she did about the whereabouts of people. I thanked her and we walked over to where the cashier stood. I asked her if she knew our wives and if she did, where were they. She eyed the grimy stars on my collar then told us that they would probably be at the hotel. I thanked her and we left.

"Did you get the impression that they disliked us in there?" Lawrence asked as soon as we had stepped out into the dusty street.

"We do look like tramps with these clothes," I replied. "They probably think that we are deserters or something. After all, no one knows that the war has ended."

"You are probably right. Do you think that we should have told them?"

"If we do not find our wives in the hotel we will go back and get a bite to eat and tell them then. My leg is killing me so I think that it would be best if we sat in there and waited for the girls instead of running all over this place looking for them. For all we know, they might be at Phanta 1 or 2."

The hotel was across the street from the restaurant so it did not take us long to get there. I asked the lady behind the counter if Debra Carson or Penny Parker were staying there. She looked through her book and told us that they were, but they had left their rooms earlier that day. I told Lawrence that they would show up at the restaurant sooner or later and so we went over there to wait on them. When we arrived, the same young waitress met us and took us to a booth.

"Miss," I said when we had sat down. "You can spread the word that the war is over. We destroyed the super base earlier this morning and negotiated surrender from the United Nations. The rest of the men will be home later tonight."

"Sure," she said. "I will also tell everyone that the United Nations is going to step out of the way and that the United States is going to send more transports with more colonists and everything will be back like it was before the United Nations came to this place." It took me a minute to realize that she did not believe me about the war being over.

"Young lady," I said in a stern voice. "The war is over and you can choose to not believe me, but I am telling the truth."

She looked me right in the eye and said, "You are right, I do not believe you. You are probably a deserter anyway. Look at your clothes!"

I shook my head. "My name is Flynn Carson. I am the commander-in-chief of the Republic's armed forces and the former pilot of the Phanta 1 mission. I was the seventh human to step foot on this planet. As for my uniform, we have been living in the wilderness for the last month. Did you not see the wounded soldiers that we brought in here two hours ago?"

Everyone in the restaurant was silent and listening to our conversation. "Sir, I have been in this stuffy building since eight o'clock this morning. They could have had a parade go by today and I would not have known it. What can I get you to eat?" I was dumbfounded at the young lady's skepticism. I ordered a double cheeseburger and fries and stepped out of the booth to address the crowd while Lawrence gave the girl his order.

"Folks, you can believe this information or call me a liar if you want, but it is the gospel truth," I began. "This morning the Republic of Phantasmagoria destroyed the super base of the United Nations. We bombed it, and then we stormed it on the ground. You may have seen the wounded soldiers come in on the planes and be taken over to the doctor's office. After we destroyed the base, we contacted Earth and the United Nations agreed to surrender. They have agreed to stay out of the way of other nations and their dealings with Phantasmagoria. The United Nations also agreed to pay war damages to the Republic of Phantasmagoria to the amount of six billion United States dollars. They are going to pay this in different metals which will be used for the benefit of the people. Thank you all for listening." I started to sit down. "Oh, I almost forgot, the rest of the soldiers are supposed to be returning tonight." The ladies

Phantasmagoria

cheered when I said this and then they went back to their conversations. The skeptical waitress appeared to be slightly embarrassed as she took our orders to the kitchen.

"Flynn!"

I recognized the voice and sprang out of my seat before I even could look up. I threw my arms around Debra and squeezed her tightly. I could her tears in her voice as she told me how much she had missed me and how much she loved me. I tuned my head and smelled the sweet scent of her blonde hair. I wished that the moment would never end. After being away for a month straight and seeing her only a couple of times in the two months before that, it felt good to know that I would not have to leave her for a long time.

"The war is over," I said. "The United Nations surrendered. We can now be together again, and this time forever."

"Thank the Good Lord," she said. "I never want to be away from you again."

"Don't worry, I will find us a nice house near a lake like I promised you and we will go there and never leave. I know just the place." We kissed, and then sat down in the booth. Curtis cried from the ground where Debra had set him in his baby carrier. She pulled him out and handed him to me. He giggled and waved at Lawrence who was sitting across the table with Penny. The waitress came back and took Debra's and Penny's orders. She also told us that the meals would be on the house in honor of the occasion. We thanked her and then started to catch up on things that had happened while we were gone. Debra told us that not much had happened since we ran the United Nations out of the towns they had occupied and destroyed their local bases. The only major thing that had happened was someone had opened a bank in Seeone and started issuing gold currency. The idea had caught on in other villages and soon almost all towns had a bank or two. Lawrence and I did not tell them much about the war because most of the details were not very pleasant at all. I told the girls about my wound. Debra wanted to examine it right there, but I assured her that the guys had treated it well and I would be just fine.

After the meal, we decided to go to the hotel and play Monopoly. I did not realize just how much I had missed Debra until I was able to see her again. Being away from her for so long had given me a new view of our relationship. I truly knew that the Lord had made us one flesh and it would be unnatural for us to be separate. I decided that once things settled down I would suggest the lake I had stopped by with that vision for the site for another village. I had a dream that needed to come true.

∞

"Oh Flynn, the view is so beautiful!" I had taken Debra back to the Doolittle range to finish the picnic that had been disturbed by the arrival of the United Nations. We had spent an hour cruising in and out of canyons. I then flew us to a plateau somewhere within the canyons and landed. We had just climbed out of the aircraft when Debra made her exclamation. I looked around. The sun was shining in the air, heating it to around one hundred and ten degrees, but it was enjoyable in the low humidity of that region. A slight breeze was blowing and that helped make the temperature more comfortable. All around us for miles were canyons. This was a very beautiful planet, I thought as we sat down on a large boulder.

"Flynn, do you have a pocket knife on you?" my beautiful wife asked. I did, so I pulled it out and handed it to her. It was a Swiss Army knife, so she opened the screwdriver on it and started scratching it into the soft sandstone that we sat on. This is what she wrote:

Flynn and Debra Carson, Day 130, Year 2. Earth date: July 20th, 2040. We have made it.

She closed the knife and handed it back to me. I pocketed it and then put my arm around her shoulder. We sat there staring out over the canyons for a while. It felt good just being together. I felt that only now was I truly living. My life on Earth had just been to make my life on Phantasmagoria seem better and brighter. On Earth I had never been so happy and Phantasmagoria had taught me so much. I never really knew the true meaning of love until I was separated from the love of my life for a period of time. I never really knew the true meaning of faith and trust until I saw the way my soldiers trusted me. I never really knew the true meaning of devotion and loyalty until I saw that wounded soldier use his life's energy to salute me. I never really knew the true meaning of sympathy and compassion until I knelt by that soldier, begging the Lord to spare his life. On the other hand, I never really knew the true meaning of war until I saw living men instantly turn to ash in a swirling hell of burning thermite. I never really knew the true meaning of desperation until I saw those men standing by the manhole openings to their underground shelter, desperately firing at us while their own bodies were ripped apart by our machine gun bullets and grenades. I thanked the Lord for all of my experiences, both the good and the bad, and for the lessons that they had taught me.

After a while my appetite broke up my thoughts and I climbed down from the rock. Debra followed and I got the picnic basket out of the plane. I carried it over to the shady side of the rock and we sat down to eat. Debra had made turkey sandwiches and cherry pie. I looked at the meat which had been grown in an incubator and at the cherries which had been grown in a greenhouse and marveled at how technology had helped us to survive on this planet from a different solar system. But technology was not the only thing to thank for our survival. Survival was a thing that came from the heart. A person had to have the will to survive; technology could not do it for him. The colonists on Phantasmagoria had come with the will to survive, even though some of them did not quite know what that would involve at the time. On the other hand, we had done more than survive, we had thrived and lived. We had built a social infrastructure that was still developing and that infrastructure had helped us through our war of independence. We had also started to form our own culture. We had new styles of architecture, new styles of pottery, as well as a new way of life. All of this started from the people. Technology could not have done it alone.

"What are you thinking about?" Debra asked.

I smiled at her. "I was just thinking about this world and the people." I picked up another piece of pie and began to eat it. "I was thinking about what this planet will be like in the future."

"Oh," she said.

"Someday this planet will be crowded with people. They will have a national park at these canyons and they will be nothing but a tourist attraction."

"Don't be silly. There will be people who live here. Those canyons are a great place for farming and mining. Look at that fertile stretch of land over there by the river."

"Yes, but people will have forgotten what it took to get here. They will forget about the blood and sweat that was shed in order to establish the Republic. We will be nothing but history book characters."

She took my hands in hers. "We will be remembered for the ones who fought against tyranny and gave freedom to the future generations."

"We will be remembered like George Washington and the rest of the founding fathers; looked at as heroes by those who support freedom and looked at as terrorists by those who desire to have the government give them a living."

"There will always be people like that, Flynn. There is nothing that you can do about it anyway. Why are you being so negative? That is not like you."

"I am not being negative, I am thinking about what has happened to America and I am trying to figure out a way to combat it. There has to be a way."

Phantasmagoria

"The constitution that the captains wrote will be a good start. However, people will find a way around it eventually if they are desperate enough. The only thing you can do is educate the next generation, and that starts with Curtis. You must train him to fight for freedom in the same way that you did. You must teach him to value freedom. You must also train him to train his future kids. That is all that one person can do."

"You're right. That is all that I need to do, actually." We sat together in silence for a spell. Finally Debra broke the silence with a startling remark.

"Flynn, did you know that the people want to elect you as president?"

I looked at her shocked. "No I did not know that. How do you know?"

"Everyone is talking about it. They figure that you would be the best candidate for the position after the way you led the military to victory. Wouldn't that be so neat? You could be the first president of Phantasmagoria!"

I thought about it for a minute. It was tempting because of the fame and importance of the position, but that was the only thing that tempted me. "No Debra, I will not run."

"Why not?" she asked, totally shocked at my reply.

"I would not run for office because that would not be in the best interest of my family or Phantasmagoria. We need to find a place and settle down to live out our lives. If they want to elect a good man, then they should elect Lawrence. He is the best choice by far. But I need to raise my son and stay with my wife."

She grinned with that cute grin of hers. "I was kind of hoping that you would say that," she admitted.

"You were?"

"Yes. You have already done more than your civic duty. If you want more politics, then run for the town council. I am glad you made that decision."

"You little sneak," I said, poking her nose.

We packed up the picnic supplies and I carried them to the plane. We climbed in and took off. I decided to fly north until we hit the northern most edge of the canyon range. I looked down at the canyons below us. I felt so free up there in the sky. The wind blew past us, tugging at my fresh, new clothes. I grabbed Debra's hand and shoved the joystick to the side, making us do a barrel roll. She squealed as we turned over, and then slapped me for catching her off guard when we leveled out. I laughed, and then did a loop. It felt great to be alive.

∞

I looked at the workers that were building the house. They raised the ceramic walls up and attached them at the corners. They slid the floor panels in and then they put the roof panels on. They installed the doors and windows. They put stairs up leading to the porch. I put my arm around Debra's waist and we stood there together looking at our new house. Finally, about ten o'clock in the morning they finished installing all of the little details and told us that we could move in. Our furniture would be arriving in an hour, so they had perfect timing. We waited for the workmen to leave before going inside. I caught a hold of Debra just before she went in, picked her up, and carried her across the threshold. I set her down inside. We did not say word, but just stared around the house, mouths agape in surprise. The interior of the house was beautiful. The walls were a light green color and the ceiling was white. The floor had a light pinkish-red tiled appearance. The walls had lots of large windows which let in lots of bright sunshine. The air conditioning system kept the air in the house at a pleasant seventy-six degrees.

I went into the kitchen. A huge glass window gave a beautiful view of Sweet Home Lake. I had purchased all of the property from our house down to the lake so that we would not lose the view. Debra joined me. "It is almost too good to be true," she said in a quiet voice. I simply nodded.

A little bit later the furniture guys came with three trucks full of furniture. They unloaded all of the furniture and let Debra pick out what she wanted. The workers carried her purchases inside the house while I paid the boss. We shook hands then helped carry furniture into the house. By the time we were done, Debra had transformed the inside of the house into the inside of a home. After the furniture men left, I surveyed the place with approval. Debra filled some vases with water and went out to the lake to pick flowers. I walked over and sat down in the huge, cushy, reclining chair. I leaned back and propped my feet up. I then quickly got up and walked over to the counter, pulled my money out of my pocket. I put it back in and headed to the market to get some groceries.

The streets were full of people working. The village was being built in one day and so everyone was hustling. The ceramic houses only took an hour or two to assemble, but a transport was due in from Earth at five so the guys were trying to get as many houses built before then as possible. I walked past the greenhouses to the open-air market. I went from booth to booth selecting vegetables and meats. Finally, my arms were full and my pocket was empty so I headed back home. Home. It felt so good to finally have a place that I could call home. I looked at the small cottage before entering. Yes, it was a home; there was no doubt about that. I went inside and put the groceries inside the refrigerator.

Debra came in with her arms full of flowers and put bouquets all through the house. I followed her into the living room where she was putting flowers on the mantle above the fireplace. I pulled a pretty pink one out of the vase, snapped part of the stem off and tucked it into her hair above her ear. She giggled and told me to go get Curtis from Penny who was babysitting him while we got settled into our new home.

As I went down the street to the courthouse, I wondered how Curtis's friends would treat him when he told them that he had been babysat by the first lady. Some would believe him and some would not. They would most likely be quite a bit jealous of him. I went into the courthouse and thanked Penny for her time.

About four-thirty that afternoon I headed out to the field where the transport would be landing. This was the first transport to this region of the planet. So far all of them had landed to the south. Even though Seeone had been named the capitol of the Republic, people had wanted to move south and populate those regions instead. I was thankful that I had been able to use my influence to get them to make Sweet Home and get a transport to land here. There were about fifty people waiting on the field for the transport to land with their new neighbors. Finally we heard the boom that signaled the new arrival hitting the atmosphere. I looked to the west and saw the glowing flash of light. The fiery object got closer and seemed like it was aimed right for us, but I knew that it would land a quarter of a mile away. The drogue chutes opened and lowered the transport to the ground. It made a nice soft landing on the turf and sat there, cooling off from the intense heat generated by hitting an atmosphere at such a high speed as was necessary for interplanetary space travel. An hour later the hatch opened and the gangway was lowered. The crowd on the ground started to approach the transport. I walked up and stood about ten feet away from the bottom of the staircase. The new colonists began to exit the craft that had been their home for the past six weeks and walk out into their new, permanent home. I watched as people filed by. Someday I would know most of them if not all of them by name, but today I was just content to watch the expressions on their faces as they looked upon the planet for the first time. Most of them were shyly and nervously jabbering to their family and friends as they stared at their new neighbors. It was like a party where the new kid in town shows up and no one quite knows just what to say to them.

I was standing there about to leave when I heard a voice that I thought I recognized. I looked up to see a Hispanic man walking down the gangway with his wife and small son. "Jerry!" I yelled and

Phantasmagoria

climbed up the stairs as fast as I could. When my leg healed from the bullet wound it had left me with a small limp and stiffness that made it hard to walk up stairs. Jerry and I met with a handshake and a bro-hug halfway up the stairs. He introduced me to Javier, his son. I walked with them to the village where we found their house and got them settled in. I invited them to come over to our house for supper since there was so much that we needed to talk about. I was thrilled that my best friend had come to the planet and had happened to make his home in the same village as me.

When Debra saw Jerry and Cheryl, she almost cried. She gave each of them a hug and told them to sit down. I had hoped that she had enough food prepared and she did. We sat down at the table with bowls of chili to talk about everything that had happened since we had left Earth. It turns out that president Lee was not the good, constitution supporting person he had claimed to be during his campaign. He made multiple policies that completely violated the Rights of the American citizens. Some people were just fine with this and perfectly content to sit in their living rooms playing video games while their fat welfare check came in each week. But then there were those who saw the promise of freedom across the gulf of space and decided that America was too far gone to take back. These people were the ones who paid for the welfare checks that were given out so freely by a broke government. Jerry had decided that it was time to go so he bought tickets for Phantasmagoria and here they were. They planned to start farming and raise cows. The meat incubators could provide meat for everyone that tasted good, but people wanted real meat from real animals.

Later that night, Jerry and I were sitting out on the porch in rocking chairs. We each had a cold glass of lemonade and a plate of doughnuts. Javier and Curtis played with toy trucks in the yard. I looked at the row of small houses which had been assembled that day.

"How has life been here on Phantasmagoria?" Jerry asked, breaking the silence. "I mean other than the War for Independence, which you probably do not want to think about."

"On the contrary, my friend, the War of Independence is something that I never want to stop thinking about. Sure, the scenes of battle are not pleasant, but if we forget about the pain and suffering that it requires to earn freedom, then we will forget about freedom. I cannot let that happen. As far as life on this planet goes, I feel so free when I walk around. It is such a simple way of life here. I believe that that is the way we are supposed to live. I do not think that humans were ever intended to live with the hurrying that goes along with life on Earth."

"That is the first thing I noticed when I stepped off of the transport; a feeling of peace," Jerry reflected. "It is so peaceful here."

I looked around at the setting sun, the empty street, and the grass covered hills beyond the town. I looked at Earth hanging just above the sinking sun with its moon tagging along behind it. I felt the cool evening breeze blow across the porch and heard the sound of laughter coming from the kitchen. "Yes," I agreed. "I think that one of the greatest gifts that the Lord has given us recently is Phantasmagoria."

Epilogue

I steered the small boat over the clear waters of Sweet Home Lake towards the dock. The electric motor made a quiet, humming noise as it propelled me through the water. A small bird flew over my head with a minnow wriggling in its beak. I looked at my day's catch of fish in the bottom of my boat. It had been a nice catch and would provide enough money to buy groceries for the rest of the week.

I steered the boat against the pier, climbed out and tied the boat securely to the pylon. I grabbed my stringer of fish and walked into the small, peaceful village. People were walking around from place to place shopping, gossiping, and visiting. I strolled into the center of town and the turned down the small street that I lived off of. It would have been quicker if I had just walked along the shoreline because my house was on lakefront property, but I enjoyed walking through the village and seeing the people. I nodded a polite greeting to some women that passed me going about their shopping.

I approached my small, cozy cottage and was about to open the door when my beautiful wife opened it. She stepped out onto the porch and gave me a kiss before I could say anything. She backed away and took the stringer of fish from me. "I will get these cleaned up and down to the market for you," she said, sweetly. "Go ahead and pick a nice watermelon to have with lunch."

I walked off the porch and around the corner of the house to the small vegetable garden there. I selected a very ripe, juicy looking melon and cut its stem with my knife. I picked up the melon and walked back to my front door. I was about to step up on the porch when I got a familiar feeling, kind of like I had already done this at some point in my life. I turned around and sure enough, a young man with my build was walking away down the street. He walked smoothly and without a limp though. I grinned and walked into the house. "Debra," I addressed the blonde at the sink. "Where is Curtis?"

"He is down by the lake playing with his ball," she said. "You had better put the melon in the refrigerator so that it's nice and cold for him when he returns."

I grinned to myself as I put the plump melon on the refrigerator shelf. My dream had come true. It had come true in a way that I never could have imagined. I went into the living room and sat in my recliner. Yep, the Lord had dealt me a pretty good hand when he put me on Phantasmagoria. I picked up the glass of lemonade that Debra always set on the table for me when I got home from the lake and took a sip. Yes, life was sweet.

The End

Made in the USA
Lexington, KY
10 August 2014